THE VIEW
FROM 4-SPACE

ANTONY C. SUTTON

1st Edition by FTIR Publications, 1999

Copyright © 1999 Antony Sutton

Reprint Edition, Dauphin Publications, 2015

ISBN: 978-1-939438-37-9

dauphin publications

Table of Contents

Foreword

Dr. Sutton apprised me that the paradigm extends even to the field of "Energy." He sent me his first edition book, "The View from 4-Space," substantiating that The Control Paradigm extends to the very fabric of the 4th Dimension. With the help of Thomas Eugene Bearden I learned that most inventors are denied funding if their inventions provide significant 'overunity' aspects inherent within them. Funding is only provided to favored companies having access within The Control Paradigm. Many do not realize that significant suppression techniques are applied to real overunity technologies that reach fulfillment protocols.

Overunity is real but it resides within the dimension of 4-Space, necessitating the knowledge of Quantum Mechanics, and the physics of Dirac, Prigogine, and an understanding that you are now delving into the realm of the Creator Himself. The work of Bearden and Bedini have interesting side effects known as 'precursor engineering' dynamics, in which physical reality itself is affected. Dirac discovering the truth inherent within Quantum Mechanics was immediately suppressed and was told that he could not divulge specific components of his discovery. His discovery of negative energy and the reversal of the time wave, verified the initial completed Unified Field Theory of James Clerk Maxwell.

Leslie R. Pastor
New Energy Congress
Member

Preface

We live in the materialist physical world of three dimensions (or 3-space). What we observe with our senses is what is real. If it cannot be observed and measured, it does not exist. This is a sure comfortable world, the ability to observe and measure gives us certainty.

With this certainty, we have in the last 200 years built an industrial society with technologies based on measurable energies and devices controlled by tight all-embracing and not to be challenged physical "laws."

Except that we have a problem. These physical 3-space "laws" do not embrace or explain every observation. There are numerous exceptions, anomalies and inconsistencies to the laws. Moreover, our 3-space technology ignores even the possibility that there may exist forces and activities we cannot measure or observe. In brief, our reality is incomplete and if incomplete, our human laws are also incomplete.

So we ask, is there another reality? Another dimension which originates these anomalies and inconsistencies that are not explainable in 3-space terms? That our puzzling anomalies are actually the effects of an unseen causal source.

This book explores technological anomalies to find they indeed form a paradigm that cannot be explained in physical materialist terms. This alternate paradigm cannot be explained with our 3-space terms, but has its own set of laws and conditions. This we call 4-space and the view from 4-space is awe-inspiring, to say the least.

So, starting from 3-space effects, which are truly anomalies in 3-space, we go to an equally real, but not observable, 4-space to explore for causes. We find new forces, new explanations and a more powerful challenging dimension.

This dimension has the seeds of, and is the basis of, future technology.

Chapter 1
Structure of the New Science Paradigm

For more than a decade we have explored generally unknown aspects of technology rejected by the physical materialist three dimensional (3-space) system in which we live.

Our exploration has ranged from ancient technologies, and "ancient mysteries," alchemy, anthroposophic science, the work of Dr. Wilhelm Reich, and especially the numerous discoveries abandoned, rejected and sometimes even suppressed in the last two hundred years because they do not conform to conventional 3-space thinking and rules. (1)

Our guide has been more or less seat-of-the-pants practicality....if it works technically then it's worth pursuing from the sociological-economic viewpoint whatever the opinion or reaction of conventional science. *Does it work*? That is the only guide to inclusion.

Readers may wonder what links such diverse technologies as homeopathy, N-machines, gyroscopes, and remote viewing in our mind, or what prompts us to look at alchemy in the frame of cold fusion or place special emphasis on the biological effects of non-ionizing radiation.

Intuitively, we look for common factors. Theoretical linkages that together explain an entire paradigm. A paradigm cannot be a collection of ad hoc technologies—it must have a common theoretical explanation. Here we outline in an elementary way our perception of the theoretical links between many apparently disconnected, dissimilar and out-of-the-mainstream technologies. In each case we find that the New Science technology kicks in, or an effect emerges, at the point where certain nonphysical parameters can be identified.

In the N-machine, an over unity electrical generator, rotation of a cylindrical magnet at 7,000 rpm induces over unity (i.e., more out than in). In homeopathy, dilution of the substance past Avogadro's Limit kicks in the "second curve," the subtle energy phenomena or increased efficacy which led to the discovery that water has a memory. In cold fusion deuterium loading past the 80% point kicks in excess heat generation. In etheric weather engineering Constable uses rotating dimension tubes to "draw" orgone energy. The orgonomy movement uses similar water grounded tubes. In meditation an inner world is entered beyond the physical world. In remote viewing the Second State is entered by rotating out of the physical body.

The common factor is the new dimension entered or "gated" by the technology (thus we call it gated 4-space). This new dimension has its own characteristics, not those of our observable 3-space physical world, but specific to 4-space dimensions.

The gating can be achieved by specific geometrical motion. N-machines, gyroscopes and Constable etheric translators use rotation as the gating mechanism. Homeopathy uses rhythmic succession, i.e., the ten count in successive waves. The geometry of breathing is related to meditative states. Cold fusion requires a precise manner of loading, fast and heavy. In remote viewing entry to the ether is gained by 180 degree rotation of the physical self

Physical laws only apply to our physical dimension of 3-space. Other laws presently unknown or vaguely understood apply to gated 4-space. Those scientists who deride the etheric phenomena are confined to 3-space thinking. It is also conceivable that some or many original experiments have been forced into physical "laws" where the etheric may more closely approximate a useful explanation. One example that comes to mind is the experimental work of N.A. Kozyrev at the Pulkovo Observatory in the former Soviet Union on the properties of time. (2) Kozyrev notes certain phenomena such as loss of weight while rotating in gyroscopes and attributed this to physical explanations (i.e., the rotation of the earth). Whereas for political reasons Kozyrev was forbidden to explore the possibility of 4-space factors (the Soviet Union, of course, being the ultimate in materialism and rejection of spirituality).

For 20th century observers steeped in materialism it is extraordinarily difficult to visualize how thought, intent and purpose are interwoven with 4-space laws. This is sheer nonsense in 3-space. So deeply is this bias merged with the fabric of a 20th century scientist that the reaction to 4-space phenomena is emotional and distressful. One cannot influence a diesel engine by thought. In 4-space thought is an energy, and while it cannot influence a diesel engine it can have other influences, and this possibility escapes materialist science.

In fact, it may be that intent and purpose are operating energies in 4-space. This supposition can explain the hit-and-miss results of certain phenomena. Many years ago we asked former U.S. Air Force General George Keegan (one-time head of U.S. Air Force Intelligence) about the capabilities of "psychic warfare." Answer? "It's too erratic." This is also the conclusion of CIA and DIA work on 4-space procedures in remote viewing (although we question the CIA/DIA public pronouncements).

Our interest is "what causes the erratic nature?" One cannot, like Randi the Magician and John Maddox, former editor of *Nature*, throw out the baby with the bathwater and dismiss all psychic phenomena. One has to determine the operating laws. This will also challenge the many fraudulent operators using "psychic phenomena" for material gain.

Our first exposure to the role of intent was when we met Trevor James Constable, an unusual scientist in the classic understanding of the word, merchant marine officer and developer of the etheric engineering version of weather control. In the early 1980s we appeared on the Ray Breem Show (KABC) in Los Angeles to discuss etheric weather engineering. One of the call-ins was a gentleman who said that Department of Commerce had tested Constable etheric translators and could not make them work. Constable's answer was that the "state of mind of the operator" is part of the process. State of mind, intent and purpose, influence the outcome.

Certainly those working to understand the new paradigm or building its technology are mostly not motivated by power, control or greed, the basic elements of today's materialist societies. A search for truth appears to be a common factor in most cases.

For the moment, we will assume that gated 4-space energy, the etheric in varying forms, will not kick in for purposes inconsistent with the conditions and laws of 4-space. In brief, a Rockefeller or a Pentagon cannot monopolize these technologies, they cannot be used for self-aggrandizement, and it is conceivable that they cannot be used for destructive purposes. We will see later that diamagnetism, the life force, is constructive and life-building, that right-turning spirals have similar properties. So does centripetal movement.

To be sure, these are shattering assumptions, especially for someone like this author who was raised in the enterprise, profit-making, materialistic mode. But facts are facts.

"Our initial understanding is that the New Science and its associated technology are based on gating 4-space energy in various forms from the etheric 4-space dimension and its processes are governed by different rules, and certainly the rules of 3-space do not apply."

<u>This explains the extraordinary irrational hostility aroused by these gated 4-space technologies.</u> In fact, unless science keeps its head, we may be entering another period of the Inquisition.

The FDA jailed Dr. Wilhelm Reich and burned his papers for investigating life energy. Professor Huizenga, of University of Rochester and a key advisor to the Department of Energy, tried to kill cold fusion and led a pack of scientists screaming "fraudulent" and "impossible."

Richard Milton (3) put it this way:

"Something nasty has crept into public life in the past 50 years and tainted the whole liberal humanistic intellectual tradition, like a dead rat under the floor boards."

Milton calls it *"hard authoritarian cynicism that springs to spit ridicule on anyone with the audacity to question any of its beliefs."*

Of course, these phenomena are impossible for 3-space laws and thinking. They are just not governed by 3-space laws, that's all.

These hostile critics sense, however, the extraordinary challenge to their privileges and acquisitions posed by the New Science. (4)

Notes

(1) Our work in technology goes back to the 1950s. See Antony C. Sutton, *Western Technology and Soviet Economic Development* (Hoover Institution Stanford University, Stanford, 1968-1974, in three volumes).

(2) N.A. Kozyrev. Translated title, "Possibility of experimental study of the properties of time," May 1968. Available from *Future Technology Intelligence Report* Special Reports, P.O. Box 2903, Sacramento, CA 95812

(3) Richard Milton, *Forbidden Science: Suppressed Research That Could Change Our World*, (Fourth Estate, London, 1994).

(4) All this is beautifully described by the late Thomas Kuhn in *Structure of Scientific Revolutions* (University of Chicago, Chicago, 1962).

Chapter 2

The Conventional View of the Future

The conventional mechanistic futurist view of the 21st century technological paradigm as expressed in journals such as *Scientific American, Popular Mechanics*, and by various futurist organizations and reflected in books and films on "the future," is fundamentally different from the emerging paradigm.

Even Arthur C. Clarke, the best known and most capable of futurists, has only modified his outlook in the last few years. In 1993, Clarke addressed the Pacific Area Senior Officer Logistics Seminar at the Hilton Hotel in Colombo, Sri Lanka. In the address, Clarke unabashedly accepted cold fusion, which has been almost unanimously rejected by the physics establishment:

"It is beyond serious dispute any more than anomalous amounts of energy are being produced from hydrogen by some unknown reaction..."

Clarke proceeded to discuss the implications of this discovery. At the same time, Clarke sent a letter to Vice President Gore repeating his conclusion and adding,

"Clearly, no effort should be spared to resolve this matter speedily by supporting scientists who are obtaining results and perhaps discouraging those who have been obstructing them." (1)

The divergence between the conventional view of the future and a future based on unlimited energy is simply because conventional thinking is based on extrapolation of present 3-space technology. Molecular electronics, thinking machines, cloning, artificial intelligence, nanotechnology all developed in the materialist paradigm are *assumed* to be elements in any future paradigm.

Following the 18th century renaissance, a materialist scientific upsurge generated technology that vastly improved the material well-being of mankind in the Western world. Steam engines, electricity, internal combustion engines, atomic power replaced animal power for work in the fields and new industrial revolution factories generated products for the masses.

Michael Faraday, a bookbinder's apprentice who never attended a University undertook a remarkable series of experiments in the 1830s and 1840s which became modem electrical engineering. Faradays original notebooks are even today in 1998 fascinating reading. One has to marvel at Faraday's ability to visualize the potential in simple observations with the barest of equipment and little prior knowledge. Faraday was followed by Maxwell, Helmholtz, Lenz, Edison, Marconi, and above all, Nikola Tesla.

Many of Faraday's observations, especially on diamagnetism, were not developed by those who followed and have been ignored by the quest to know more and more about less and less. When modem scientists criticize those who discover new fundamental properties outside their field the put down comment is "Oh, he's not a scientist." They need to remember that Faraday was a bookbinder's apprentice and by modem standards could not even hold an electrician's job. Most of today's scientists are actually highly skilled technicians, and when we look for example at diamagnetism we shall see that even those who followed Faraday have no ability to understand the full sweep of his work. How many electrical engineers have explored diamagnetism?

In transportation, the Watt steam engine became the railroad revolution to replace the horse and canal barge. Then came trucks and aircraft from Ford and the Wright brothers. Medicine evolved from barber ship surgery, bleeding leeches and herbs to modem drug and surgery techniques. Food is no longer coarse bread and beer for the masses, but refined package products and bioengineered products.

Many more names, like Bell, Sikorsky, Whittle, Pasteur, gave us our remarkably successful materialist paradigm. So obviously, all-powerful and all-consuming science and technology have ignored many other insights, concepts, and anomalies.

Among the most significant of ideas ignored is the scientific work of Goethe; in fact, almost all recognize Goethe as poet and writer, not a scientific genius. Only in recent decades under the leadership of physicist David Bohm has a reevaluation of Goethe science evolved.

Goethe rejected the mechanical causality of scientific orthodoxy and was scorned for his emphasis on nature. Yet as Henri Bortroft points out, the mechanism of causality cannot be applied to some phenomenon as color where Goethe undertook highly penetrating experimental work. (2) Goethe's theory of color is unknown, yet is more accurate observation than Isaac Newton's.

Perhaps even more unknown is the work of Rudolf Steiner and his secretary Guenther Wachsmuth. (3)

The genius Nikola Tesla is barely known. His invention of alternating current and hundreds of other devices is buried by the name of his commercially successful competitor, Edison. Flying wing developer Burnelli is unknown, yet he produced safer and more economic designs than Douglas and Boeing.

The future as seen today is therefore based on selective past developments, those that were commercially successful in the past and controlled by a narrow group of robber baron capitalists in the late 19th century. This is an inefficient and fragile basis for extrapolation.

Above all, science has come to depend on physical observation and measurement. If you can't see it, you can't measure it, and therefore it does not exist. We have taken the god of materialism so far that even consciousness is interpreted as a chemical reaction.

This remarkably successful materialist system has now, however, frozen its store of knowledge without explaining even such fundamentals as life energy, emotional experiences and thought processes. The ultimate absurdity of the materialist paradigm is that its fundamental building blocks, atoms of matter cannot be given a logical origin. "Big bang" theories explain nothing. Multibillion dollar super colliders bashing sub-atomic particles tell us nothing about the nature of matter itself.

The fatal flaw of materialism and the widely visualized future built on a materialistic basis is that it bypasses all anomalies that do not fit materialism and explain origins with assumption.

Thomas Kuhn, in *Structure of Scientific Revolutions*, touches on this flaw:

"Normal science, the puzzle solving activity, is a highly cumulative enterprise eminently successful in its aim the steady extension of the scope and precision of scientific knowledge. Yet one standard product of the scientific enterprise is missing. Normal structure does

not aim at novelties of fact or theory and when successful finds none." (4)

The Tomorrow Makers

The tomorrow makers are the ultimate human product of this mechanistic age racing to be first to "download the contents of the human mind into a computer housed within a robotic body so that we will never have to die." (5)

This group of futurists is located at Carnegie-Mellon University at the Autonomous Mobile Robot Laboratory and Massachusetts Institute of Technology in the United States. In Japan groups are located at University of Tokyo and Waseda University.

These futurists live in a world of switches. While their enthusiasm dedication and ingenuity are impressive (especially their goal of a molecular mechanical computer in a cell)...they have convinced themselves that the future of man is either out in space or a miniaturized robot.

Quite how this robot will function without life energy, without a personality, without self-awareness of "I Am" is not explained and cannot be explained. This is the ultimate Man is God fantasy. To be sure futurists do discuss the concept of God (recorded in Grant Fjermedal) but God is dismissed as "the Force" with the admission that "you don't get to be Number One by copying unless you are copying God." (6)

The temptation to dismiss this futurist rambling as a fantasy created by pizza-chomping nerds has to be tempered. The tomorrow makers are an excellent example of the old adage of scientists knowing more and more about less and less. Futurists have taken mechanical technology to its logical ultimate without even passing reference to the possibility of new discovery.

Expansive and expensive arguments are made about using nanotechnology to repair cells and how this will revolutionize medicine. This extrapolation ignores a central fact—that many diseases, of which cancer is one, are an effect of the industrial age. At some point we have to look to prevention. How to adjust nutrition and environment to eliminate cancer.

Nanotechnology would be a goldmine for medical technology and hospitals but a disaster for those millions with cancer. The benefits of the 3-space materialist age came with costs; for example, cancer (which barely existed 150 years ago), stress diseases, tolerance to drug medicine, and so on.

To suggest nanotechnology as a cure-all solution is a blind alley. We need to prevent the problem, not accept the problem as insolvable.

Undoubtedly, computers and miniaturization will be valuable as facilitating technologies in the 21st century, but impressive as the achievements might be they are but building blocks within a more life-oriented paradigm.

Rigidity in Modern Science

It is this closed minded, narrowly focused, rigid, rule-bound thinking that makes modem science of increasingly limited value while absorbing more resources for less benefit.

The great Max Planck made the discerning statement, "new and unsuspected phenomena are, however repeatedly, uncovered by scientific research." (7)

A Max Planck, like other prominent pathbreakers, could not accept this truism simply because they were pathbreakers. They had no investment that demanded protection. They built the paradigm.

Today's scientists are largely technicians and defenders of the mythology which sustains their technological abilities. They are not builders of a paradigm. They have a lifetime invested in a stock of knowledge, and God forbid it be challenged.

How Wrong are the Futurists?

It was after we spend a decade exploring the technological potential of New Science that we discovered the literature of futurism. Surprisingly, there is little common ground between futurist prognostications and our assessment of what is actually emerging today; alongside an advanced electronic system.

Even Arthur C. Clarke, the granddaddy of futurists and by far the most perceptive and far-ranging who understands the significance of technology, missed, until recently, the emerging technologies we include below. Clarke does make some approximations but with vague meanings:

"The age of cheap energy is over; the age of free energy is 50 years ahead."

Clarke pinpoints solar and nuclear energy and, not until a few years ago, cold fusion or N-machines or vacuum energy as the energy of the future.

A well-accepted book, *What Futurists Believe*, published by the World Future Society, has succinct statements and summaries of positions held by a dozen leading members of the craft, including A.C. Clarke, Peter Drucker, Daniel Bell and Kenneth Boulding.

One observation is striking—not a single key influence or technology mentioned here is cited by any of these experts (except A.C. Clarke's statement in 1993). <u>Futurists are stuck solidly in 20st century thinking.</u>

The extraordinary work of Dr. Wilhelm Reich is omitted, so is Nikola Tesla. Not even glancing mention of etheric weather engineering, zero point energy, implosion vortex technology, cold fusion (announced in 1989, the same year Coates' book was published), fuel cells, life energy, vibrational medicine, cymatics, Burnelli flying wing design in aircraft . . . all with enormous future potential.

These futurists follow the same route as the former Soviet Union, they extrapolate the known, not explore the unknown. This rigidity in the Soviet Union led to collapse, the system could not make rational technological progress. The futurists will not give us collapse, just a horrendous waste of resources.

And there is no excuse. John White's *Future Science* was published in 1977 and is entirely consistent with our findings here. In other words, New Science has an identifiable continuing track which can be analyzed. If futurists don't find the track they are incompetent or intellectually rigid.

Moreover, academics like Dr. William Tiller (chairman of the Materials Science Department at Stanford University) were publishing revolutionary findings in the 1970s.

This extraordinary blindness by futurists is parallel to our findings in *Western Technology and Soviet Economic Development* (three volumes, published at Hoover Institution Stanford University, 1968-1974). Our findings on Soviet technology were directly contrary to all academic assessments, including the State Department and the Central Intelligence Agency. These institutions had analysts locked into a specific viewpoint and mind set. However, both the Federal Bureau of Investigation and the U.S. military had noted aspects of our findings without probing the full implications.

In projecting future technology as with Soviet technology the prevailing orthodoxy has been sidetracked from the truth by peer pressure (i.e., don't challenge the community consensus) and peer review (don't publish what is not generally accepted), compounded by an unwillingness to explore anomalies and exceptions.

Static Government Forecasting

Government forecasting and academic economic discussion are often way off target because they fail to consider technological change.

In some states, California for example, the State itself may not purchase or contract for undeveloped or new technology. California firms have forced through legislation that makes it impossible and, indeed, illegal for the State to use "unproven methods." Unproven by whom? By those who would suffer financially. In other words, vested interest has forced through legal protection for a specific manner of technological operation.

Thus, while Singapore, Malaysia, Namibia, Eritrea, Greece and Israel have officially and successfully used either DeMeo cloud busting or Constable etheric weather engineering, the same process cannot even be considered in California. And as long as "official science" says the process is impossible there is no chance of media coverage beyond the "fraud" and "quack" approach. Therefore, no chance of public opinion forcing change.

Back in 1975, the U.S. National Research Council undertook a comprehensive study of the prospective energy economy for the coming years and tried to forecast 10-15 years ahead by technological segment. (8)

NRC has the authority, expertise and enough funding to encourage, or at least not discourage, more efficient energy sources. While we have no knowledge of the extent of NRC exposure to New Science we do know that in 1975 the Government of Canada was aware of the new paradigm, and in fact the Canadian Senate encouraged the collection of data. In the U.S. the potential was ignored. Members of the NRC panel appointed to consider the 25 year projection all, without exception, had a stake in the orthodox contemporary energy paradigm as it existed in 1975. The appointees came from atomic, gas, oil, coal, and engineering construction firms that specialized in building units for these conventional technologies (i.e., Bechtel, Stone and Webster, EPRI, Electric Power Research Institute), while academics from Cornell and University of California both known for their anti-new energy adherents, filled out the panel. In brief, not the slightest representation from the New Science, at that time. In later years EPRI did undertake some successful research in cold fusion, otherwise no link at all to the future!

However, in 1975 when the NRC panel was making its projections there were hundreds of individuals and groups at work, using their own funds and working on N-machines, magnetic motors, implosion technology, heat exchange systems and fuel cells. These were rejected out of hand by orthodoxy on the basis that science already knew everything and certain known laws governed all physical phenomena.

Consequently, it was easy for anyone to dismiss these new ideas as "perpetual motion machines" developed by "kooks" and "crackpots." This is a way of disposing of arguments and facts that cannot be disposed of in any other manner. Indeed there were some impractical ideas among the surfacing methods...but there were also many that deserved investigation and possibly development. Instead, the Patent Office, under influence of the U.S. Department of Energy, dismissed applications that did not conform to scientific principle; i.e., an assumption that all principles were known. The famous Newman machine was rejected for years until Newman got Congress to force the Patent Office to grant a patent.

In 1975 there were in fact reports from Stanford University, as we shall see below, that confirmed over unity and stressed its "tremendous implications". (9) Peer pressure and intellectual laziness suppressed this information and inventor De Palma was fired from MIT. Over in Switzerland and Germany work had begun on over unity magnetic devices. In India and Japan individual scientists were also exploring these devices. Yet the NRC panel was so fixated by man-made laws of physics, and incidentally, their own pecuniary interests, that they could not envisage even a remote possibility that dramatic advances in energy technology were down the road.

Why not? Because to recognize even the possibility of anomaly was a threat to a lifetime intellectual investment, their power, influence and material well-being. These advances were sufficiently powerful to threaten the "system."

On the other hand, in Ottawa, Canada, in the mid-1970s, a senior Canadian Government official named Dr. Andrew Michrowski was picking up these same reports of over unity—from the U.S. as well as from Canada. Not only the Canadian Senate, but also Prime Minister Pierre Trudeau and then-Governor General Romeo Le Blanc became involved.

Compare this to the United States where entrenched political and corporate powers control "the system" so effectively that even today, 20 years later, no Congressional hearings have been held and any New Science development is shouted down by pack journalism and a rigid scientific establishment.

Cold fusion, for example, moved overseas to France and Japan simply because the hostile atmosphere in the U.S. made development almost impossible. In the end, it was the U.S. Navy that said "we made a mistake" and after a decade, cold fusion became discretely respectable.

Only in new medical ideas did the establishment given even an inch, and it *was* an inch. After citizen protests over AMA-FDA suppression of preventative medicine, the U.S. Department of Health was instructed by Congress to devote some resources to these new ideas, especially for cancer treatment. This resulted in a magnificent effort...a staff of four in an Office of Alternative Medicine! Their final report, *Alternative Medicine: Expanding Medical Horizons,* contained so many disclaimers as to guarantee little impact. (10)

Dr. Michrowski's comments are worth quoting:

"An enigmatic situation has emerged. Over the years, a number of credible, clean energy systems have been researched and quite a few have reached refined state of development—certainly much more than was even dreamt possible twenty years ago. Nevertheless, very few aspects of the new clean energy technology are reaching the industrial or commercial mainstream. It is tragic that the solid and multifaceted achievements of Dr. Yull Brown and his Browns Gas, Brandson Roy Thomson's inertial propulsion drive, John Hutchinson, Nikola Tesla's untapped discoveries research in vacuum energy, longitudinal (scalar) waves and advanced magnetic motors are languishing in a maze of disinterest or simple impotency."

"The politics leading to this logjam are arabesque. They challenge even the ordinary alternative technology and "soft energy path" initiatives worldwide."

"Ultimately, the democratization and the political demands of an informed public can assure the general population access to clean energy." (11)

In conclusion, the future paradigm is here in embryo form, and will be described in detail below. The paradigm is entirely ignored in conventional forecasts for the future. Why? Because the new paradigm offends the pocket book and intellectual stock in trade of those who comprise the contemporary paradigm community.

The gut truth is that there is profit in smog, profit in pollution, profit in drought, profit in floods, even profit in inefficient systems of energy generation.

This is not an argument against profit, which is an essential mechanism for an enterprise economy so long as resources remain scarce. But making profit through the political system is wrong, and powerful economic interests are using the political process to inhibit free movement of resources into these new technologies.

Notes

(1) *Cold Fusion*, Vol. 1, No. 1.
(2) Henri Bortoft, *The Wholeness of Nature* (Goethe's Way Towards a Science of Conscious Participation in Nature), New York, Lindisfarne, 1996, p. 255.
(3) Guenther Wachsmuth, *The Etheric Formative Forces in Cosmos, Earth and Man*, Anthroposophic Press, New York, 1932.
(4) Kuhn Op. Cit., p. 52
(5) Grant Fjermedal, *The Tomorrow Makers*, MacMillan, New York, 1986, p. ix.
(6) Op. cit., p. 240
(7) Max Planck, *Scientific Autobiography and Other Papers*, New York, 1949, pp. 33-4.
(8) National Academy of Sciences, U.S. Energy Supply Prospects to 2010, Washington, D.C., 1979.
(9) Over unity = over 100% efficiency. In layman's terms more power out of the device than used to operate the device. Laws of thermodynamics say this is impossible, but see below.
(10) *Alternative Medicine: Expanding Medical Horizons.* Pace News Letter, Vol. 8 (2), p. 18.

Chapter 3

Gateway to 4-Space

The Industrial Revolution, now 200 years old, will be supplanted by another now-emerging technological revolution...the Etheric Age.

Unknown to almost everyone, including most of the scientific community, a fundamental new technological paradigm has surfaced without the support and certainly without the recognition of science, Government or the academic world.

Although individuals in Government and the academic world have made contributions, in general the work has been undertaken and positive results achieved in the face of overt and covert opposition. In fact even knowledge of the New Science has been concealed.

In the last decade or so these individuals have formed into groups, become formalized and now issue their own journals, hold conferences and have established small scale organizations.

The academic community is guilty of outrageous misbehavior to maintain its own world view, prestige and funding. The typical reaction is "it's impossible" or "I don't want to know about it".

Why are they so upset about the possibility of an advanced technology? Briefly, as we shall see below, the current system is challenged and the pecking order threatened. If any major element in the New Science is allowed to surface to public view, for example either "cloud busting" or etheric weather engineering, entire knowledge segments (i.e., in this case meteorology) and their supporting industries (flood relief, drought relief, real estate values, forest fire fighting) face overnight extinction or wrenching modification. These interests work to keep the gate to 4-space closed.

Hyperdimensions as theoretical structures are well established in mathematics, physics and esoteric systems. But technological links to a hyperdimension are currently seen as anomalies or isolated aberrations rather than ingredients of a hyperdimensional technology.

On examination this 4-space technology is not only internally consistent but operationally is far more powerful and far superior in many ways to the 3-space technology of the 19th and 20th centuries that emerged from the Industrial Revolution. Four-space promises (almost) free energy, a nonpolluting energy, reactionless drive systems (non-Newtonian), regional climate control to eliminate smog, fog, floods, drought and forest fires, and more powerful and more natural medicine.

The vast potential of 4-space explains why it has remained secret for so long. Every facet of our society, every investment, every political structure, the entire fabric of our stock of knowledge is challenged. Our contemporary scientific structure is as outdated as the 16th century scholastic doctrine of a flat earth, the divine right of kings and a wagon wheel economy which ushered in the Industrial Age.

Now this Industrial Age is about to give way to a far more advanced system, an etheric system based on energetic properties gated from 4-space. Not only is the new system economically more efficient (even scarcity of resources is challenged) but is without the adverse spillovers of pollution and destruction.

The materialist physical dimension and its accompanying subsets of activities are long past the point of diminishing returns. The new 4-space is not an extrapolation of contemporary 3-space. It is not a bigger and better conventional system as the future is illustrated in glossy magazines and TV dramas. This is fundamentally new science based on newly identified forces and new ideas although some elements were identified as far back as antiquity.

This little book introduces these 4-space technologies with a brief description of their history and potential. In many cases the technical feasibility is more or less known but the market place has still to identify the most efficient among variations.

These new systems will generate titanic struggles and pressures in society. We doubt that new systems will become dominant, or even generally recognized before 2050 or later. Our entire political-economic-scientific-academic structure will be overturned, in other words the etheric system is a body blow to the materialist system.

The far sighted will grasp the opportunities and move ahead. Reluctant power holders of the present will resist, defend the "system" and sabotage at every turn in the road.

Early Workers in 4-Space

A fascinating side view of science history is that some of these concepts were identified in the 19th century in the first years of the Industrial Revolution but then put aside and not pursued. It has taken almost 150-200 years for the potential to be realized.

Michael Faraday, Samuel Hahnemann, Johann Wolfgang von Goethe, Baron von Reichenbach, James Maxwell, John Tyndall, Nikola Tesla all lived and worked in the 19th century and made contributions to 4-space recorded in their work.

There was a contribution by the ancients, (now being revived) through the mystery schools, mathematics and sacred geometry, the use of earth energies which is more than symbolic mythology. Von Goethe is the starting point for the modem era but his scientific contribution has been overshadowed by his fame as a poet and because so little of his science work has been translated into English. Goethe understood the significance of the spiral and the role of polarity. But his most developed work is the concept of color as *an edge phenomenon*, very different from, more experimentally accurate and more realistic than Isaac Newton's theory of color (1) Essentially, Goethe had the insight of nature as a living body rather than the dead lifeless nature of materialism.

Michael Faraday is especially notable. The father of electrical engineering left the idea of a homopolar generator, ignored for 150 years until the late Bruce De Palma at MIT visualized its free energy potential and developed the N-machine ("n" for the nth possibilities) as one of a number of devices that operate over unity.

Faraday worked on diamagnetism significant in the New Science as the source of life energy. This work has been entirely overlooked although recorded in the DIARIES. Faraday in the 1830's can also be seen as the father of the diamagnetic cold suction technology formulated by Viktor

Schauberger in the 1930's and the 1940's and adopted by Germany for the flying saucer-type vehicle in 1944.

The genius of Michael Faraday has been recognized by the anthroposophic movement and Rudolf Steiner which originated the concept of a spiritual science. (2) Indeed many anthroposophic ideas reflect von Goethe and find their way in 4-space technology.

The religious writing of Swedenborg, a 19th century mystical expression of higher space was an early precursor of spiritual ideas in the context of non-Euclidian geometry. (3)

What do these early ideas suggest for our everyday world?

Simply that fossil fuels, atomic energy and systems based on heat and pressure including many 3-space alternative systems are on the way to replacement. Their problems will also be replaced. The Environmental Protection Agency and related organizations are unnecessary, the new technologies are pollution free. Neither can existing electrical generation methods meet the new standards of efficiency. A 15% efficient coal fired plant cannot compete with over unity ranging up to 300-400% efficient.

The Nature of 4-Space

For centuries devotees of religious and esoteric wisdoms have entered 4-space, a higher dimension, through the mind and religious ecstasy. Their teachings focus on human attainment of inner vision and truth through medication, psychical and ceremonial practices.

Erwin Laszlo in the *Whispering Pond* (4) guides us through this vision of science by reminding us that past civilizations and religions all have the concept of a higher dimension and that the recent Western Einsteinian view of cosmology is limited and inadequate.

More satisfying according to Laszlo is the work of the former David Bohm, basically that our misunderstanding of reality is rooted in physics but not limited to physics... the concept of the implicate order.

Others like Shiji Inomata in Japan consider the mind and thought as essential to the emerging technological system. Inomata gives equal weight to consciousness, matter and energy.

These concepts are here elevated to a new stage: the exploration of technological processes or access to a higher dimension to tap higher dimension properties. These are the subtle or etheric energies not identified or recognized in 3-space. Mind and thought become elements in the technology.

By a process of rotation or geometrical re-positioning etheric energy can be channeled from 4-space into the physical 3-space. When this concept is presented to conventional scientists, they exclaim "that's impossible" and cite the laws of conservation of matter and the laws of thermodynamics. They do *not* tell you that these laws are man-made and therefore are *hypotheses* limited to closed 3-space systems. By definition they rule out any external source such as 4-space.

Rotation opens the gate to an external source where man-made laws do not apply. So what was previously a mathematical concept now becomes the practical explanation for operational technology.

The materialist paradigm has ignored the greatest wonder of all: the power of the human mind. Today's Western scientists place themselves in the role of God, all powerful and all knowing. They ignore the simplest observation: that without the mind there is no science. This omission is the single most destructive fault of 3-space Western science.

Eastern scientists even those trained in the West are more influenced by religion and culture. They see the mind and thought as an integral part of the science. Thus Shiuji Inomata in Japan is able to visualize the role of thought in the new equations and develop the concept of "shadow energy". By contract Western science rejects the role of consciousness and pushed the research into the isolated field of mind/body studies.

Our expansion of the boundaries of science is difficult and disturbing for many, impossible for some and almost always generates intense emotional distress.

When faced with living nature and the infinite power of the mind, the typical materialist scientist disintegrates. While a few physicists like Shrodinger and Bohm were able to appreciate the necessity for infusion of Eastern ideas into Western physics, most would prefer not to even consider the mystical and spiritual.

Consequently a book like *Scientific American* editor John Horgan's *The End of Science* is appropriately subtitled *Facing the limits of knowledge in the twilight of the scientific age* (Helix 1996).

In one way, Horgan is right, the scientific era is in its twilight zone. The end of the materialist scientific age is in sight but Horgan did not see another new inspiring view...the rise of a new spiritual science.

The Book of Genesis and the New Science

The New Science has more in common with religious concepts and certainly is more in harmony with the spiritual than the materialist.

Materialism holds that mind and humans evolve from matter, that all can be explained in terms of chemicals and the physical. Life energy is not recognized at all. The assumption is that man is no more than a chemical compound while biology rejects even the concept of the life force.

While modem biology deals with the dead, we find Genesis 2:7 closer to the New Science, i.e., God breathed life into man. *The spirit of God is the life force.*

A virtually unknown Canadian radio engineer Willard B. Smith (5) has a cosmology parallel to Genesis, i.e., in the Beginning was nothing and our first task is to understand nothing or the concept of vril, the primordial black darkness out of which all arose or was created.

In modem cosmology sound is merely a phenomenon that arose with the Big Bang. In the New Science reflected in Smith and contemporary researchers like Dale Pond, sound is the three dimensional shaping mechanism and consistent with John 1:1 "In the beginning was the word" ("word" is sound). The translators could equally have used the word "sound" instead of "word". What emerges here is the possibility that future science is closer to the *literal* interpretation of the Bible than to modem materialist concepts.

The concept of sound and creation was first explored by Ernst Chladni in the late 18th century and published as *Die Akustik*. This German physicist found that every sound has a specific frequency and creates a specific pattern. This was confirmed in the late 19th century by several others including the clergyman Rev. John Andrew (6) who wrote *The Pendulograph* (1881)...an extraordinary little book.

In the late 20th century the Swiss Dr. Hans Jenny followed this early work on sound and pattern using modem electronic equipment. From this emerged the science of cymatics founded in the United States by Jeff Volk and even a system of medicine based on sound, founded by Dr. Guy Manners in England and Jonathan Goldman in the United States. Parallel to this is the extraordinary acoustics research work of Dale Pond in Oklahoma, who has linked music to creative energy and revived the unknown work of John Keely from the 19th century.

It is notable that these two centuries of work has been mostly by non-professionals or professionals working outside their field. Apparently, modem science is so tightly woven that it cannot even consider ideas outside its chosen interpretation and certainly none that may suggest a spiritual element to science. Of course, a fundamentalist interpretation of the Bible applied to science is humorous to the materialist scientist.

Hyperdimensions: The Access Route

The terms 4-space and 4-dimensions have been widely used in mathematics and esoteric literature. Our use is specific and relates only to technology, i.e., that there exists a dimension to reality that is as source of presently unidentified energy and which does not conform to the rules of physical material space (or 3-space).

Our starting point was the numerous observations and anomalies recorded over the past two centuries and related history back to ancient times and that cannot be explained in terms of modem 3-space science. Our interest is technological not theoretical. That means devices and systems that can operate and have effects within our dimension for practical purposes but with causal origin in 4-space.

Esoteric writers have only touched on this technological aspect. For example, *The Fourth Dimension* by C. Howard Hinton (7) touches on the phenomenon this way:

"To determine if we are in a four dimensional world, we must examine the phenomenon of motion in our space. If movements occur which are not explicable on the suppositions of our three dimensional mechanics we should have an indication of a possible four dimensional motion... "

Later Hinton adds:

"This form of matter I speak of a four dimensional ether and attribute to it properties approximating to those of a perfect liquid" (page 16) and concludes that:

"All attempts to visualize a fourth dimension are futile. It must be connected with a time experience in three space" (page 207)

We cannot therefore reject a system simply because it does not fit the various hypotheses advanced by orthodoxy. We believe that to reject measurable repeated results simply because they do not conform to a theoretical structure is absurd, especially if the devices have unusual value and promise.

In the process of accumulating these anomalies and devices we surmised that there had to be an explanation. The most general explanation is that these anomalies and disregarded technologies are associated with the ether, also known as the quantum vacuum, zero point energy or the neutron sea. The latter may even be an almost massless gas and which was at one time in the original periodic charts designed by Dmitri Mendelieff. (8)

There is support for this in Hinton. According to Hinton, the phenomenon of an electric current, which is found in many 4-space technologies is consistent with the hypothesis of a fourth dimension and rotation of a fluid ether. In brief *over unity electrical generation is a 4-space effect.*

This explanation for over unity is also found 80 years after Hinton in Moray B. King who portrays fourth dimension energy arising from an orthogonal electric flux.

Figure 1

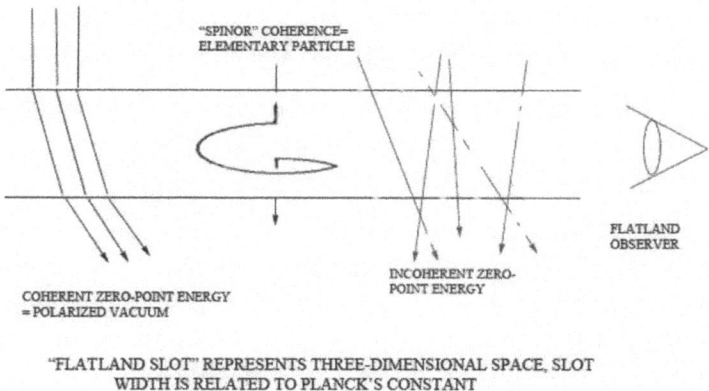

THE ZERO-POINT ENGERY MAY ARISE FROM AN ORTHOGONAL ELECTRIC FLUX FROM THE FOURTH DIMENSION

"SPINOR" COHERENCE=
ELEMENTARY PARTICLE

FLATLAND OBSERVER

INCOHERENT ZERO-POINT ENERGY

COHERENT ZERO-POINT ENERGY
= POLARIZED VACUUM

"FLATLAND SLOT" REPRESENTS THREE-DIMENSIONAL SPACE, SLOT WIDTH IS RELATED TO PLANCK'S CONSTANT

Source: Moray B. King, *Tapping the Zero Point Energy*, Paraclete Provo 1989

Figure 2

CREATING ORTHOROTATION BY BUCKING FIELDS

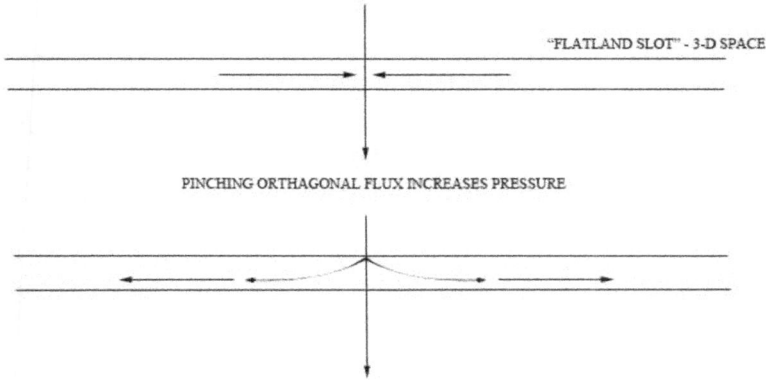

"FLATLAND SLOT" - 3-D SPACE

PINCHING ORTHAGONAL FLUX INCREASES PRESSURE

ABRUPT RELEASE CAUSES FLUX TO FOLLOW LINES OF VACUUM POLARIZATION

Our analysis of gated 4-space is illustrated in Moray B. King, *Tapping the Zero Point Energy* (9). For King, zero point energy arises as an electric flux that flows orthogonal to our 3-space. This is not the only access route or gate but King has elaborated in detail on this interpretation.

Despite physicists denials over free energy there IS a considerable discussion in the physics literature about a hyperspace which contains enormous quantities of energy. It is generally argued that this energy cannot be tapped because it is incoherent, i.e., random. The King theory is that the existence of a hyperdimension suggests certain conditions under which the ZPE flux becomes coherent.

The orthorotation required for ZPE to enter our 3-space can be created by various devices including the caduceus coil where appropriate or mirror image windings generate abruptly pulsed bucking fields. Such bucking fields are used in numerous designs of over unity motors (i.e., the Swiss ML Converter, Carrs, Sears and Newman motors).

Below we consider not only the Hinton-King gating mechanisms, but also the Constable geometric transducers, homeopathic vortices in the 9th dilution known by Benveniste and Schiff as the "second curve", rotating magnetic cylinders in the N-machines and other related devices. These and other mechanisms are gates from a 4-space cause to a 3-space effect.

Notes

(1) *Goethe: Scientific Studies* (Princeton University, Jersey, 1995). Also Heinrich O. Proskauer, *The Rediscovery of Color* (Anthroposophic Press, New York, 1986).

(2) Ernst Lehrs, *Spiritual Science, Electricity and Michael Faraday* (Rudolf Steiner Press, London, 1975). Thomas Martin, *Faraday's Diary* (Bell, London, 1932), 5 volumes.

(3) Claude Bragdon, *Four Dimensional Vistas* (Knopf, New York, 1925).

(4) Ervin Laszlo, *The Whispering Pond* (Element, Rockport, MA, 1996).

(5) Willard B. Smith, *The New Science* (Murl Smith, 1964). Reprinted by DW French, P.O. Box 2010, Sparks, NV 89432.

(6) Rev. John Andrew, *The Pendulograph* (Bell, 1881). Reprinted by Borderlands, P.O. Box 220, Bayside, CA 95524.

(7) C. Howard Hinton, *The Fourth Dimension* (Swann Sonnenschein, London, 1906). Readers interested in the

expression of tesseracts, i.e., four dimensional cubes generated by extending every part of a cube in a fourth direction at right angles to each other will find much material in Hinton.

(8) Dmitri Mendelieff, *An Attempt Towards a Chemical Conception of the Ether* (Longmans Green: London, 1904).

(9) Moray B. King, *Tapping the Zero Point Energy* (Paraclete, Provo, UT, 1989).

Chapter 4
Essential Elements of the Emerging Paradigm

Work in the new science of the etheric is worldwide, surprisingly consistent, and only sporadically published in orthodox scientific literature. Some elements as etheric weather engineering have never been mentioned at all in science literature and remain entirely unknown to the world at large.

The New Science paradigm has its own literature, including journals and a remarkable number of books. Patent literature is a valuable source. However, the U.S. Department of Defense has over 6,000 classified patents and over 750,000 secrecy orders which may not be discussed. Fortunately for us, most of these concern 3-space development, not the etheric 4-space arena.

We suspect a few U.S. classified patents include some elements of the New Science as, for example, the Townsend-Brown electro-gravitational engine and the U.S. Navy secret level project to build a unipolar generator. (1)

In general, however, the etheric bypasses 3-space advances and is itself not amenable to classification. How does DOD classify remote viewing, for example? When it originated as a civilian discovery and the skills required cannot be classified?

Journals include *Journal of Borderland Research* (founded 1945), *Speculations in Science and Technology* (England), *Pace Newsletter* (Canada, since 1975), *Science and Medical Network Newsletter* (England), *Future Technology Intelligence Report* (Since 1990), *Safe News* (Swiss Association for Free Energy in German), *New Energy*

News (since 1992), *Frontier Perspectives* (Temple University), *Space Energy Newsletter* (Florida), *Electric Spacecraft Journal* (Florida), *All Source Digest* (Washington), *Nexus* (Australia), *Cold Fusion Times* (Massachusetts), *Infinite Energy*, (New Hampshire), *Aura-Z* (Moscow, Russia in five languages), and many others.

Normally, many new journals and magazines fall by the wayside within a year. This is not the case with the New Science. We can only recall a few cases of failure in New Science, and that is because the production was too far ahead of the market. *Cold Fusion*, published by experienced publisher Wayne Green, was a superb large format, glossy, sophisticated publication but only survived one issue. Almost certainly, this was because the market at this time does not have sufficient depth to support an expensive production. (2)

There are several Japanese publications in Superscience reflecting a more open Japanese view of the Future. (3) The Editorial content varies considerably. Some, like *New Energy News*, are heavily technical and restrain from criticism of the government. Others, like *Nexus* in Australia, are openly rebellious and anti-government.

There is an extraordinary unanimity among these journals in their view of the new paradigm and its component elements. Here we have individuals scattered around the world in only haphazard contact, yet their basic view is the same: there is an etheric dimension far more powerful and interesting than our materialistic physical dimension and one which generates numerous effects in our world.

Whether one approaches the topic from the direction of energy, from medicine and biology, from music and acoustics, from weather engineering, from altered states of mind and remote activities, one ends up looking at the same etheric picture.

Essentially, the following elements comprise the paradigm at this time.

• New energy generators, mostly over unity. These can be rotating magnetic cylinders, vortex diamagnetic turbines, electrochemical devices like cold fusion with the common feature that they gate energy from 4-space;

• an associated field involves shape and geometry. The ancient ideas of earth energy linked to sacred geometry. Geometry is also intimately linked to gating some 4-space energies to 3-space. This element includes various subtle energies;

• the concept of a life energy is central to a medicine incorporating homeopathy, naturopathy, massage, herbal medicine, Chinese and Ayurvedic energy medicines. The concept of life energy can be found extensively from the Bible to Eastern religions in ancient mythology and almost everywhere *except* in the physical materialist system. The idea of a life-oriented world has been submerged by the physical non-life;

• the incorporation of consciousness into theories of matter and energy. Consciousness is not merely a brain phenomenon but a unique spiritual phenomenon including out-of-body states with its related remote phenomena;

• etheric weather engineering, or "cloud busting," in the original concept.

The New Science is not compartmentalized. All segments are related to the etheric. As all of today's researchers started *de novo*, it is no great burden to span the entire field of the new knowledge from the etheric viewpoint. We will draw examples from ancient Chaldean rotating tops to present day electrical generating rotating magnetic cylinders, i.e., from biblical archeology to electrical engineering.

Two centuries ago, von Goethe spanned all science and the arts in a situation somewhat analogous to the new paradigm. The binding core is the recognition of the etheric. In *Borderlands* journal, for example, the range of articles is from mechanical and electrical engineering to ancient art and from physiology to geology. This is not an unusual range.

The core of the paradigm is the recognition and revival of the ether or primary energy of "empty" space in its varying forms. This is an old concept generally accepted up through the end of the 19th century when the emergence of materialism drove it out and the Michelson-Morley experiments supposedly proved the ether did not exist, ignoring the scientific truism that the negative can never be fully proven.

So thoroughly has the etheric concept been eliminated that we have encountered narrowly trained "scientists" who have never heard of the ether and even confuse it with the chemical ether. This narrow training of modem scientists is a significant roadblock to understanding the etheric. These are skilled technicians, not scientists.

Back in the early 1980s we presented an informal lecture film shown to a group of engineers and hydrologists on etheric weather engineering. At one point the chairman intervened—a foreigner trained at the University of Liverpool—and said, "Wait a minute, these people don't know what the ether is...they are too young."

Today's massive scientific structure is not a positive influence for the future. These people extrapolate only what they know, and the etheric has long ago been ruled out...with one important exception.

Quantum electrodynamics requires such a concept and introduces the neutron sea in place of the ether. Space is not empty for this school. Space is a reservoir of energy. The practical question for most quantum physicists is whether it can be accessed. Orthodoxy says no, the energy is incoherent. Yet the new science technology is tapping the reservoir of space (i.e., there do exist technologies to cohere the vacuum energy).

At this point we can outline the technologies that comprise each segment of the new paradigm and how each links back to an etheric 4-space source.

Free Energy

Nothing is free. This is a misnomer. It is the fuel that is free in the new paradigm. Even if we draw energy from the limitless vacuum of space we still have to invest in capital equipment and distribution facilities. "Free energy" means that coal, oil, natural gas and atomic fuel are not needed. Fuel is either space or water.

The rotating magnetic devices in various designs are the original free energy devices. The earliest such device we have a record of is the Hans Coler coils over magnetic device built in the 1930s and recorded by British Intelligence Report *Bios 1043* in 1946. (3)Then came a variety of magnet motors known as Newman, Sears, Johnson, Monus, De Palma, Tewari and Inomata, to name a few. The source of excess power is the fourth dimension, or 4-space.

Another group is implosion cold suction devices, which also date back to pre-World War II in Austria and are known as Schauberger devices. The energy here is related to diamagnetism, also identified as a form of life energy and developed from spiral turbines. (4)

The recent cold fusion devices are electro-chemical, based on electrolysis of water using platinum/palladium rods with heavy or light water or catalytic based on palladium. The origin of the excess energy is still in discussion, but according to Moray B. King, is 4-space (see below).

Shape and Geometry

There is a group of energetic effects known as subtle energies which have both communicative and influencing effects.

Some forms are described by Mesmer and his followers. Others are termed "earth energies" and related to dowsing and ley lines. Feng shui and acupuncture are leading technologies.

The key book describing ether technology is Guenther Wachsmuth's *The Etheric Formative Forces of Cosmos, Earth and Man...*, based on the extraordinary work and insights of Rudolf Steiner. Here you will find an analysis of the ether and its component parts and the relationship to geometry and color.

This group has an extraordinary variety of forms and appearances yet to be fully explored. Life energy is related to cone and pyramid shapes, and in the East, formalized in feng shui. In etheric weather engineering dimensions and shape are significant in the rotating tubes.

The vortex and polarity are essential aspects, and sound is a form-creating device used even by NASA for levitation.

The Life Energy is Central to Medicine

What is the life force? It depends on whom you ask... The word "life" suggests the province of biology...but don't ask a biologist. Contemporary biologists concern themselves with the structure of the dead, not the essence of the living. The life force, including the source of logical, purposeful human action, should be at the very center of biological studies. It is not, and why it is not is a study in itself.

Our research uncovered almost 200 individual discoveries of the life force (6), and many more probably exist. There is an ancient tradition going back to Neolithic times through the Egyptian dynasties, still known in Tibet and more corruptly among primitive peoples that demonstrates an everyday awareness and interplay with a life force. It appears to be a massless, or nearly massless, energy that creates influences and controls life, matter and destiny.

In Western cultures the life force is known as the Holy Spirit, but materialist science shuns anything linked to the spiritual in any form.

However, what is emerging today is an appreciation that the life force is central to health, that a balanced life energy is essential to health and sickness can be avoided with a balanced life energy and to some extent cured by correcting imbalances. The limited drug surgery materialist techniques are expensive, inefficient and outdated, except for emergency trauma situations where they excel.

Mind-Matter Relationships

The emerging interest in life energy and the spiritual elements of technology have opened up new approaches to the relationship between mind and matter, including consciousness. The leading source of information in this segment is *Frontier Perspectives*, published by the Center for Frontier Sciences at Temple University. In Japan, Shiuji Inomata has elaborated a schema for the 21st Century, including consciousness as interchangeable with matter and energy. Physics and zen intertwine to become different facets of the universal. This goes a long way to explain the so-called paranormal and has been adopted in medicine as dielectric diagnosis.

Frontier Perspectives is a most productive gateway to explore the numerous organizations and publications in this segment of the New Sciences paradigm.

Etheric Weather Engineering

The first technology to evolve entirely within the framework of New Science is etheric weather engineering and "cloud busting." Beginning with the work of Dr. Wilhelm Reich, suppressed by the U.S. Food and Drug Administration, this technology today has two divisions... the DeMeo continuance of the classical Reichian "cloud buster" and the development of the Constable version in etheric weather engineering. The same basic technology has been developed in Switzerland and Italy and has been used in many smaller countries.

The primary energy (orgone) used in weather engineering is equivalent to the chemical (sound) ether of Guenther

Wachsmuth and goes back to the Greek concept of nature based on four elements plus the ether. This concept is totally foreign to 3-space science. While the ether concept lasted into the early 20th century, the idea of four elements was dead by the early 19th century.

The first cloud buster was manufactured by Dr. Wilhelm Reich, an Austrian psychologist who trained under Sigmund Freud and discovered what he termed orgone, or life energy. Reich manufactured the first cloud buster in 1940, was able to affect atmospheric orgone energy to induce rainfall, a method continued today by Dr. James DeMeo at the Orgone Biophysical Laboratories in Oregon.

This early Reichian approach using the "cloud buster" was developed by a merchant marine officer, Trever James Constable, after 1960 to incorporate the work of Rudolf Steiner and Guenther Wachsmuth. This included the four elements concept revived by Rudolf Steiner in the 1920s. From this Constable developed the geometric translator which accesses the ether using a combination of dimensions, vacuum and rotation along with electronics to track artificial storms. (5)

This division of weather work into two groups is important because while they both originate in the ether they use different approaches and objectives. DeMeo concentrates on reversing desertification, with a system reflecting Reichian psychology and sociology. Constable takes modem electronics and marries it to a Steiner-Wachsmuth interpretation of atmospheric energy flows.

Neither concept is understandable to modem meteorology. There is nothing, *absolutely nothing*, in meteorology that can explain or even understand primary (etheric) weather engineering in either the DeMeo or Constable versions.

So narrowly trained are meteorologists that they can only look at weather phenomena through physical filters. The concept of the etheric is a foreign language. This makes meteorological rejection of these methods automatic, prompt and emphatic. The response is always "impossible," usually accompanied by confusion and emotional distress.

Notes

(1) Reported in FTIR. See Appendix.

(2) Publisher Wayne Green is located at Route 70, 202 North, Peterborough, NH 03058.

(3) Copies of Japanese magazines are in Borderlands Science Foundation. See Appendix B.

(4) Callum Coats, *Living Energies* (Gateway Books, Bath, 1996).

(5) See Appendix A below and Richard Gerber, M.D.'s *Vibrational Medicine* (Bear & Co., Santa Fe, 1996).

(6) Trevor James Constable, *Loom of the Future* (Borderlands, Bayside, 1996).

Chapter 5

Present Anomaly is Future Technology

Orthodox science in almost any discipline rejects discovery and ideas inconsistent with its developed body of acceptable knowledge which forms the dominant belief structure, and ultimately, dominant mythology. Thomas Kuhn even makes absence of anomaly a measure of success:

Normal science does not aim at novelties of fact or theory, and when successful, finds none. (1)

The remarkable growth of materialist science and associated technology in the past 200 years is no exception. For all the spirited, self-satisfied talk about scientific freedom, the scientific literature is a device to control just as much as to circulate information acceptable to the contemporary paradigm. Scientific literature is always peer reviewed and the peer review process, with very few exceptions, weeds out discovery inconsistent with established dogma.

To quote Kuhn again:

"In developing any science the first paradigm is generally assumed to account for most of the observation and experiments." (2)

Journal editors are the gatekeepers. They guard the internal consistency of the stock of knowledge and by selection of peer reviewers can to a great extent control who enters the gate and who is rejected.

Some editors go to extraordinary lengths to maintain doctrinal purity. *Nature*, the long established international journal under former editor John Maddox, once sent a team to Paris to investigate the work of a respected French scientist, Dr. Jacques Benveniste.

Dr. Benveniste had discovered that water has a memory (3) ...definitely not part of the physical paradigm, although the latest findings in quantum electrodynamics suggest that water "may retain and release electromagnetic information that it has acquired in some way or another." (4)

This Benveniste finding gave credence to homeopathy, and *Nature* is heavily dependent on pharmaceutical company advertising. Editor Maddox ensured an unfavorable report for Dr. Benveniste by selecting an investigative team comprised of himself, Randi the magician and a fraud investigator from the United States. About as biased as one can get.

The Maddox bias is overt. In most cases, the editor sends submissions to referees, and by selecting referees exercises control almost covertly.

The treatment of cold fusion is typical. There were very few exceptions to universal physics rejection of the Pons and Fleischmann discovery. Only a couple of journals, e.g., *The Journal of Electrochemistry and Journal of Fusion*, gave any space to Stanley Pons and Martin Fleischmann and their initial article in 1989. The peer review dam is not completely watertight.

However, for the greater part it is a massive, if subtle, barrier to new discovery, and this was typical of the Pons-Fleischmann discovery. Not a single media science writer was positive. They all cited "the laws of physics" to prove cold fusion impossible.

The barrier problem begins with graduate students who reflect the prevailing mythology, write theses, dissertations, and blue books consistent with acceptable views. These efforts are guided and approved by academic examiners, not for originality but to cautiously push forward the frontier of the acceptable sufficiently to appear "original," by adding a package of facts or experimentation to support, but not challenge, the paradigm.

Graduate research efforts are directed to miniscule confirmations of a narrowly defined discipline, i.e., biotechnology (but not bioenergy), inorganic chemistry (but not transmutation), Soviet studies (but not Soviet technical dependence on the West), electrical engineering (but not over unity homopolar generators), and so on.

Newly degreed graduate students in these narrowly defined and internally consistent disciplines (anomalies are not allowed) become junior professors and research scientists. They usually lack the wide knowledge, understanding and incentive to make truly venturesome explorations at the frontier. In brief they are *technicians*, not scientists. They got this far by reflecting the accepted body of knowledge. The original thinkers never even entered graduate school, or had the ability to dichotomize their thinking.

Furthermore, lacking original curiosity and discouraged in graduate training, they usually lack the perception and judgment necessary for evaluation of anomalies. To quote Kuhn again:

"Further development therefore ordinarily calls for the construction of elaborate equipment, the development of an esoteric vocabulary and skills, and a refinement of concepts that increasingly lessens their resemblance to their usual common sense prototypes. That professionalization leads ...to an immense restriction of the scientists' vision and to a considerable resistance to

paradigm change. The science has become increasingly rigid." (5)

This rigidity is today the barrier to acceptance of the New Science. This is why the Federal Government, staffed almost completely with conventional scientists, is unable to perceive these new paradigm effects except when brought unmistakably home in the form of unwelcome events.

The prevailing paradigm is always protected from ideas offensive to the acceptable and comfortable. And Federal Financing of science guarantees the old will continue and the new rejected.

Anomalies will continue to be rejected as erroneous observations, and above all, prevailing principles (so-called laws) will not be challenged. This is Kuhn's rigidity in practice.

From time to time a persistent investigator breaks through the orthodox barrier...and the orthodox reaction is deafening with hostility and unrelenting in efforts to squash the anomaly and the unfortunate discoverer.

The Massachusetts Institute of Technology and California Institute of Technology were loud with screams of opposition in March 1989 when Fleischmann and Pons announced cold fusion. One would expect talented scientists to receive the announcement with applause and intense curiosity. Over unity on a table top! But this is not what scientists saw in cold fusion. It visualized collapse of the laws of thermodynamics...and disappearing federal finance.

It took the U.S. Navy to admit the truth in 1993: "We made a mistake." The U.S. Navy had its China Lake station repeat the cold fusion experiments and they came out consistently positive. But in 1997 the Congress was still pumping billions into hot fusion and ignoring cold fusion, a far cheaper process, with proven over-unity.

Less well known is the ejection of Bruce de Palma from MIT when he pursued the N-effect, developed from the Faraday unipolar generator, with unwelcome tenacity. At the time, De Palma was instructor at MIT. He was soon ejected. Why? The N-effect is an anomaly and challenges the laws of thermodynamics.

This author found the same hostile reactions in the UCLA Department of Economics in 1958-1963 and later at Stanford University, when exploring the origins of Soviet technology using a novel theory of technical transfers. The prevailing orthodox position was "the Soviets have their own self-developed technology." Yet from this author's own industrial experience, we knew this was untrue and we made this view plain in graduate seminars at UCLA. We were ejected from UCLA in 1963.

The absurdity here is that the key faculty members behind the ejection, Dr. Armen Alchian and Dr. Jack Hirschleifer, were also working at Rand Corporation in Santa Monica on Soviet strategy, which required accurate assessments of Soviet technology. It wasn't until 1983, twenty years later, that CIA Director William Casey admitted that "we [i.e., CIA] made a mistake." UCLA even today has not admitted it made a mistake. The mistake? They completely missed Soviet dependence on Western technology.

This is a good point to explore the story of our three volume *Western Technology and Soviet Economic Development* (6) in more detail because it demonstrates the extraordinary efforts required to convince the orthodox dominant group of the existence of an anomaly, and why we are so certain that Kuhn is correct when he argues that the orthodox "does not admit any novelty of fact." It also demonstrates the disastrous twists in government policy that result from orthodox rigidity, an institutional blindness that costs lives.

The way the U.S. perceived the Soviet Union during the Cold War has become known as the "perceptions problem." The CIA and the State Department were emphatic, the Soviet Union was a *real* super power with its own self-developed technology and a threat to the Western world. This of course suited the corporate ambitions of the military industrial complex. This was the prevailing viewpoint reflected in all economic development textbooks, in the media, in military planning and the export control laws. *It was the prevailing mythology*. This mythology generated giant contracts and stupendous profits. Yet it was truly a mythology.

We entered UCLA Graduate School in 1958 after a multi-country, multi-industry experience with industrial technology in Europe, Canada, Mexico, and the United States. Not once in ten years had we encountered any Soviet design technology and we knew that the Soviets were massive importers of the most advanced technology from the West. In our experience the accepted paradigm was absurd...the Soviet Union was actually built by the West and was dependent upon the West for technology. This reality of dependency was concealed by a clever propaganda image that became the dominant reality for the West.

At UCLA this view was dominant, i.e., "everyone knew" that the Soviets were far ahead of the U.S. in technology. No discussion was allowed, as this was verifiable truth, according to UCLA pundits. It was impossible to submit the contrary reality as a doctoral thesis though massive evidence existed and even in the UCLA Main Library!

After three years at UCLA, we were ejected from the doctoral program. A sympathetic member of the Department, Dr. Dudley Pegrum, found a position for this author at California State University, Los Angeles, as assistant professor of economics. At Cal State we encountered the same hostility and viewpoint: "Soviets have their own self-developed technology, everyone knows that..." But we survived five years before denial of tenure.

However, quietly, we began in 1965 a massive self-financed study, a technical analysis of Soviet industry. For five years we said little and completed the first volume of the three-volume study. The data were not difficult to obtain...the Soviets had convinced THEMSELVES they had a self-generated indigenous technology and the Glavlit censors allowed export of numerous training and equipment manuals and handbooks under the mistaken impression they were needed to maintain Soviet exports of this equipment. These manuals were undeniable evidence of dependence when compared to Western models and designs.

Then came one of those fortuitous events that cannot be explained, but broke the barrier of orthodoxy. We sent the manuscript unsolicited to Henry Regnery Company in Chicago. Two weeks later we had a return letter from Henry Regnery himself to the effect that the work was "very important but not commercial." Regnery advised we send the manuscript to Hoover Institution at Stanford University.

Again, by accident, the manuscript arrived on the desk of Alan Belmont, Assistant Director for Administration at Hoover Institution, and the one man in the entire United States best able to judge its accuracy. Belmont had just retired to Hoover from FBI where he was Assistant Director for Domestic Intelligence... and knew firsthand the efforts by the Soviets to acquire U.S. technology through espionage.

Belmont sent the manuscript to émigré Russian engineers, those who had worked in the Soviet plants using the technology we described. This route was probably the only way the manuscript could break through the paradigm barrier.

This was the breakthrough that Bruce De Palma never found. Pons and Fleishmann had to go to France with Japanese financing to work on their "novelty of fact," as Kuhn describes the barrier phenomenon. Yuli Brown, discussed below, found acceptance and award in Australia but rejection in the United States.

By 1968 we were at Hoover Institution working on the remaining two volumes. All three were published between 1968 and 1974 but not without major struggles within Hoover Institution, notably with hostility from former CIA personnel.

By the end of 1974 we had three volumes in print. But the fourth volume (on military aspects) led to expulsion from Hoover Institution in a battle led by CIA personnel trying to cover the tracks of inadequate analysis in official Washington.

The fourth volume surfaced the military assistance provided by the technical transfers and which were not suspected at all in Washington. We did not know at the time that these volumes had also surfaced the so-called "perceptions problem" that U.S. intelligence had been suckered by Soviet propaganda. The CIA wanted the military implications concealed to protect their image.

This author was the unfortunate messenger.

The Soviet Union collapsed within 20 years and the world now knows that the system was a technological hoax and an economic disaster. In 1982, CIA Director William Casey admitted that their estimate of dependence was wrong but obliquely suggested that CIA itself had discovered the dependence! In fact, it was our self-financed work going back to the 1950s that surfaced the dependence while CIA reaction to the work was to eliminate the messenger. We have no doubt that history will surface other similar, and maybe some worse, examples of the rigid structure of orthodoxy and its extraordinary cost.

The Emergence of Anomaly

Famous physicist Max Planck once commented: "New and unsuspected phenomena are, however repeatedly, uncovered in scientific research." They are almost always uncovered outside the purview and influence of academic rigidity by talented and independent individuals who can see beyond the dogma. At this time sufficient anomalies have accumulated with confirming research and validation to create the core of an entirely different technological paradigm. *This is a global phenomenon.* And even when presented in varying formats (philosophy, mathematics, and science) there is a consistent common core. That is: present anomaly is the key to future technology, or in other words, a 3-space anomaly is an effect from 4-space.

Independent researchers in all fields more far-sighted and curious than paradigm adherents have seized on the anomalies, confirmed that they are genuine and in many cases actually moved the scientific anomaly to development of working technology, almost unknown.

The peer-review literature has refused publicity, so there has emerged an extensive underground literature. These early publications have been followed by global conferences, now probably in the hundreds. There is the International Cold Fusion Conference, the International Conference on New Ideas in Natural Science with some long-standing conferences like the 22nd Annual U.S. Psychotronics Association Conference held in Columbus Ohio in July 1996.

The N-Machine Anomaly

Direct extraction of electrical energy from space uses the N-effect achieved by rotating magnetic fields and first noted by Michael Faraday in December 1831.

This concept remained dormant until the 1960s when Bruce De Palma, then an instructor at MIT, built a working model which generated over-unity. This was later repeated by Paramahamsa Tewari of the Atomic Energy Board in India and Shiuji Inomata in Japan, who built a beautifully-engineered model. These researchers reported over-unity efficiency and have since improved their models.

After ejection from MIT, De Palma went to the Sunburst Community, Santa Barbara, Calif., where in the 1980s he built more sophisticated N-machines. At this point the development follows a pattern that should be followed by more inventors...De Palma took his machine to Stanford University and arranged for Dr. Robert Kincheloe, professor emeritus of electrical engineering at Stanford, to test and evaluate.

Understandably, Dr. Kincheloe was skeptical. The De Palma machine offended the most basic laws of physics, a gigantic anomaly. The impact was sufficient to have De Palma ejected from MIT. Unless the reader has personally faced these situations, the dilemma facing both De Palma and Kincheloe can hardly be comprehended. They were challenging the paradigm.

However, Kincheloe undertook the task and reported on the N-machine. After testing and evaluation, he concluded:

"De Palma may have been right in that there is indeed a situation here whereby energy is being obtained from a previously unknown and unexplained source. This is a conclusion that most scientists and engineers would reject out-of-hand as being a violation of accepted laws of physics, and if true, has incredible implications." (7)

One can almost sense the struggle going on within Kincheloe. A professor emeritus at Stanford with a lifetime of belief in certain inviolable laws, taught to generations of students and assumed in laboratory experimentation...all this going out the window.

Conceivably, Kincheloe had talked with Dr. William Tiller, also at Stanford (Chairman of the Materials Science Department). Tiller is prominently mentioned in John White's *Future Science* and has developed extremely interesting frameworks for understanding the New Science.(8)

The irony of it!

Back in 1831, Michael Faraday's notebooks recorded routine work on rotating magnetic fields. This 1831 experimentation is still not explained by modem electrical engineering. Rigid theory excluded this Faraday work for over 150 years. In 1986, along came Kincheloe to see the "incredible implications."

Nothing changed from 1830 to 1986 except scientific attitudes. Man has clamped iron laws on phenomena incompletely and inadequately understood. Generations of students had been forced to apply these inadequate laws and any student who challenges the laws will never graduate. Congress and alumni pour funds into this artificial structure and call it "science"!

If Bruce De Palma had been less persevering against the MIT priesthood and Dr. Kincheloe less honest and dispassionate then Michael Faraday's observations could have gone another century as a quaint anomaly without significance...and the academic priesthood continued its reign uninterrupted.

Yuli Brown and the Atomic Waste Fiasco

A Bulgarian inventor living in Australia in the 1970s developed an extraordinary technology, Browns Gas, a patented hydrogen/oxygen (2:1 ratio) gas. This gas has remarkable properties. With water as fuel it can generate a safe gas which can reduce radio activity. A demonstration was arranged to show this reduction. Cobalt 60 was treated and resulted in a drop in Geiger counter readings from 1000 counts to 40, i.e., about 0.04% of the original. Today the DOE refuses to release the Geiger counter readings on the grounds this is private (!) information.

The Anomaly of Life

One anomaly is so vast, so basic, that its very existence is the signature of a paradigm with built-in decay.

Biology's fundamental should be...what is life? What is life energy? Leading to the question, what is self-awareness? What is consciousness? These questions rather than form the basis of biology have been shifted into a separate field of mind/body studies. That life is more than chemistry and physics has been by passed.

This anomaly is so far reaching that to ignore it displays a fundamental defect in the contemporary scientific mind. In *Engines of Creation*, K. Eric Drexler poses the question:

"Is there some special magic about life essential to making molecular machinery work?" (11)

To this question, Dexler replies:

"...biologists have abandoned the special magic idea because they have found chemical and physical explanations for every living cell yet studied including their motion, growth and reproduction. Indeed, this knowledge is the very foundation of biotechnology."

Biology may well have found "explanations" for biological processes but they cannot replicate the unique phenomenon of life itself. Hypothesizing about "cell repair" and "working protein machines" may explain for 20th century rigidly indoctrinated scientists who ignore what is not convenient but that is not enough for others.

The self-aware purposeful action of humans has not been created and cannot be created. Even cloning starts with a single cell of a living body. Life energy and spirit are totally ignored by mechanistic biology and mechanistic psychology.

Can man play God and create life? For Drexler and the mechanical scientists who follow Drexler the answer is "yes, it is coming." In their anxiety to prove this they brush aside a host of evidence to the contrary. In 1977, John White published *Future Science* (12) with a list of 100 independent discoveries of life energy, that "magic of life" all ignored by the conventional paradigm. Our own work expanded this list to almost 200 independent discoveries, each with its own name because conventional science refuses to publish the evidence and so researchers have no adequate prior history of what has gone before their own work.

Anomaly is Open to Fraud

The nagging problem with 3-space anomaly is that observations and conclusions can be biased, experiments can be rigged and scientific misconduct passed off as true experimentation.

Man is not perfect and pressures to win peer acceptance or just plain make a buck are temptations for the unscrupulous and the weak. This demands a skeptical approach to anomaly.

On the other hand skepticism can be used as a tool to dissuade and dismiss true anomalous observations which reflect an unknown truth from 4-space. There is a gray area where skepticism is a vital necessity for truthful advance and skepticism is also a tool to dismiss uncomfortable facts.

There is an organization of skeptics with local branches and literature which provides a cross check on anomaly. We have already cited the three member team assembled by *Nature* to investigate the work of Dr. Jacques Benveniste on memory of water. These skeptics can perform a useful function, although the Benveniste investigation was a classic example of proving what you set out to find.

If the skeptical investigation is honest and not biased and still comes up positive then the anomaly should be accepted. Skepticism, however, cannot become a belief structure. It then becomes fraud. The everything new is a false approach, destroys the potential for a cross check on anomaly.

For example Michael Epstein of National Capital Area Skeptics wrote in *Journal of Scientific Exploration*, "Does investigating the anomalous fall under the protection of academic freedom or does it cross the border into scientific misconduct?"

Epstein raises a valuable point but unfortunately it is the ultra-skeptics who have committed the misconduct. For example, Epstein writes, "I do not agree with Dr. Bockris' theories, particularly those dealing with elemental transmutation by electrolysis or biological mechanisms. In my opinion, much, if not all, can be explained by contamination and bad analytical chemistry."

Here Epstein the skeptic wanders into the very irrationality he criticizes. Epstein becomes the Inquisition. By denying at the start transmutation, including the living and ever-present example of biological transmutation, Epstein displays a theological, not a scientific, attitude. Elemental transmutations are not always fraud, as we see in the extraordinary work of C. Louis Kervan in France. (13)

Consequently, while these 4-space effects can be fraud, and certainly have been in certain cases, there is a legitimate area of inquiry backed by solid research which demands further exploration.

Notes

(1) Kuhn op. cit. p. 52

(2) Op. cit. p. 64

(3) Michael Schiff, *The Memory of Water* (Thorson London, 1994).

(4) Giuliano Preparate, *QED Coherence in Matter* (World, River Edge, NJ, 1995)

(5) Kuhn op. cit. p. 216

(6) Antony C. Sutton, *Western Technology and Soviet Economic Development* (Hoover Institution, Stanford 1968-1974 in three volumes. Still in print in 1997). Also, *The Best Enemy Money Can Buy* (Liberty House Press, Billings, MT, 1988).

(7) See Bruce De Palma, *On the Possibility of the Extraction of Electricity Directly From Space in Space Technology*, Vol. 13, No. 4, 1990, p. 283.

(8) See "The positive and negative space/time frames on conjugative systems" and "New Fields, new laws," by William A. Tiller in *Future Science*, op. cit.

(9) Planetary Association for Clean Energy (PACE), Ottawa, Canada.

(10) See *Future Technology Intelligence Report*, Feb. 1997.

(11) *Engines of Creation*, p. 17.

(12) John White and Stanley Krippner, *Future Science* (Doubleday New York 1977).

(13) C. Louis Kervran, *Biological Transmutations* (Beekman, New York, 1971).

Chapter 6

Rotation, Vortices, and Vibration

Rotation, or the "spin thing", as Professor Eric Laithwaite of University of London used to call rotation, has a key role in 4-space as a gating mechanism and is identified in a variety of technologies, some already in use.

Several are not thought of as "technology" in 3-space terms but fit the characteristics of 4-space technologies. Rotation and vortices are common attributes and sound is a creative mechanism.

For example, precognition is a disputed technique in 3-space but is understandable as a 4-space technique where, time, space and communication follow different rules.

The ancient Chaldeans used rotating tops for divination (pre-cognition) and if precognition derives from the subtle energy and conditions we ascribe to 4-space then Chaldean use of spinning tops not only suggests that in some areas the ancients were way ahead of us but might provide clues for further knowledge. (1)

The magical ritual of the Chaldeans was oriented to a higher dimension but described in the crude terms of ancient times i.e., mythological symbolic Gods and demons. Remove the crude symbolism, examine the mechanical procedures and you have a picture of a higher dimension with elementary attributes comparable with the quantum view of today.

According to Hans Levy (2) Chaldean oracles were instructed to operate with the Hercare top, a golden disk which was whirled or spun. When the disk was made to spin inwards (centripetal) the gods were being called, when spun outwards (centrifugal) they were being set loose. Later we shall see that the Schauberger technologies depend on centripetal action, not the centrifugal action of the current thermodynamic cycle.

The Schauberger vortical biological vacuum is centripetal (an inward cycle) rather than centrifugal (outwards). Interestingly the Chaldeans used this method to induce favorable weather, an ancient precursor to Constable etheric techniques using vortical gates (Chapter Nine below). In general although the Chaldean ritual was crude and immersed in symbology it does bear a similarity to the gating devices described here. Rotation is also linked to another subtle energy, the life force, or "chi." The Tibetan lamas have a tradition, to achieve longevity by daily clockwise rotation exercises. (3) Sufi whirling dervishes of the Saharan desert use rotation to achieve a trance like state, an altered dimension.

We find partial rotation, a 180° shift, in other technologies—Robert Monroe describes this in *Journeys Out of the Body* (Anchor, New York, 1971) as a means to shift the physical body into the Second Body and Locale II. Orthorotation of an electric flux is used by Moray King to gate electricity from the fourth dimension. (See below)

More down to earth in present 20th Century technology: the gyroscope has 4-space properties when precessing (spinning). According to Laithwaite, De Palma, Kelly, Baumgardner and others experimentation shows that Newton's Laws of Motion only apply to motion in straight lines where there is no change in the rate of acceleration.

A gyroscope is essentially a wheel mounted on a universal joint so that its axis may lie in any direction. While orthodox scientists claim that all gyroscopic properties can be explained by Newton's Laws of Motion this is hardly so. For example:

(1)   If you weigh a gyroscope on a precision balance and weigh it again when spinning on a vertical axis, it will always weigh less when spinning. Newton does not cover this case.

(2)   A gyroscope precessing on a tower on a horizontal plane will not topple or fall. It defies the pull of gravity. A stationary gyroscope cannot remain in the horizontal position without toppling. (4)

(3)   In the 1970's Bruce De Palma former MIT physics instructor experimented with rotating balls to observe the interaction of gravitational and inertial forces. Newton's laws do not distinguish between rotating and non-rotating objects. Neither did Einstein. In De Palma experiments repeated elsewhere, a rotating ball falls faster than a non-rotating ball. In other words rotation affects the mechanical properties of an object.

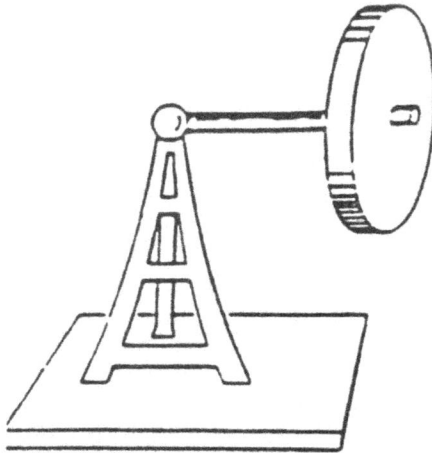

These 4-space characteristics are today used in 3-space technology without their actions being fully understood. Accelerometers, automatic pilots, gyrocompasses, gyroscopic ship stabilizers all make use of these unexplained properties. In brief we already have a few 4-space technologies proven, useful and operating.

De Palmas conclusions are worth noting:

"The momentous fact is that there is no special interaction between rotation and gravity. The behaviour of rotating objects is explained simply on the addition of free energy to whatever motion the rotating object is making. The spinning object goes higher and falls faster than the identical non-rotating control."

In brief, one can access the 4-space world of free energy or "this addition of free energy" in several forms by rotation.

The reaction of MIT to De Palma has to be noted. Honest science would record these results and draw conclusions for theory and new technology. To the contrary MIT ejected De Palma from his instructor's position. He then went to the Sunburst Community in Santa Barbara, not an academic institution, where he developed a practical over unity N-machine, an electrical generating device. In other words, a closed community not concerned with technology was able to recognize and support a technology with greater accuracy and honest response than the Massachusetts Institute of Technology which is the recipient of massive U.S. Government funds and respect. We will see later that MIT repeated this dishonest activity with regard to cold fusion. Further the author can personally confirm this suppressive activity a few years before De Palma at Hoover Institution, Stanford University.

De Palma was unable to fight the academic establishment on its own ground and later went to Australia and New Zealand.

We have no doubt that numerous similar cases will sooner or later surface. In the meantime be assured that Thomas Kuhn was right when he argued that orthodoxy will not tolerate any challenge to its dominance...they can never be wrong. We have no doubt that would be students may read this book and be concerned for their own future. Our suggestion is to keep away, from the so called "prestigious establishments" and look carefully at smaller almost unknown establishments with open-minded faculty members, or try to bypass establishment-state institutions. At some point we anticipate private voluntary communities will be established.

Rotating Magnetic Cylinders

A class of over unity devices known as N-machines has been developed which generate electricity by high speed rotation of magnetic cylinders.

As we have already noted Bruce De Palma developed the first N-machine based partly on Michael Faraday's observations in 1831. This was followed by a closed path model developed by Adam Trombly and Joseph Kahn. Later, a model was built by Paramahamsa Tewari, Chief Engineer at the Tarapore Atomic Power Station in India.

Another successful model has been built by Shiuji Inomata in Japan.

The Methernitha Community in Switzerland built a different type of space energy generator, called an M-L converter. This 200 watt model, an electrostatic/magnetic device, with similarities to a Wimshurst generator was built in the mid 1980's by member Paul Bauman and inspected by numerous outside engineers.

In a paper presented to the 26th IECEC Conference in Boston Inomata comments, "In the N-machine both electrical charge and energy in electrical power are thought to be extracted from the vacuum...considered as a balanced sea of both positive and negative shadow energy of infinite depth."

Tewari has formulated a theoretical explanation for the claim that the origin of the additional power is from the vacuum of space.

In Tewari's words:

"The machine is essentially a conducting cylindrical magnet rotated at high speed around the axis with magnetic field parallel to the axis. Since there is no relative motion between the magnetic field and the conducting cylinder, the appearance of DC voltage between the shaft and the periphery and consequently generation of power cannot be due to Faradays law of electromagnetic induction." (5)

Tewari was able to achieve a power ratio of 33.8 close to the De Palma achievement of 28.2. A Trombley-Kahn closed path version achieved similar power ratios.

For those who are looking for countries with responsive governments the experience of Tewari is worth noting. The Indian Government has been fully supportive of his work providing logistical backup and a favorable work environment. Similarly Inomata has had support of Government—industry groups although Japan does have internal opposition to the New Science. The sad story of De Palma in the U.S. is well known. (6)

Another development down in New Zealand is interesting not only because of the status of the inventor but because of the extraordinary hostility of the (then) New Zealand Government.

The Adams Pulsed Generator was developed by Robert Adams, former Chairman of the Institute of Electrical and Electronics Engineers (New Zealand section). Adams is an electrical engineer with fifty years' experience in designing and building power stations, broadcast facilities and airport communications.

Adams built a number of permanent magnet electric motors "some of which demonstrated an electrical and mechanical efficiency over 600%.''(These run at room temperature, a point that will be noted by electrical engineers i.e. heat is the major result of hysteresis losses in conventional motors or generators).

Lucas Industries in England evaluated an Adams motors at 100% efficiency.

The political reaction in New Zealand was extraordinary. The following comes from the Australian journal Nexus:

"[Adams] has survived an attempt on his life by an individual affiliated with the New Zealand Secret Intelligence Service and the Central Intelligence Agency [of the United States] direct suppression of his invention by former (and recently deceased) Prime Minister of New Zealand Robert Muldoon... "

Subsequently, Adams released detailed information on construction (7) to thwart any further attempts at suppression by governments and out of control intelligence agencies.

Vortex Devices

The vortex has been recognized and useful application made in fluid dynamics with devices like vortex flow control valves vortex pressure amplifiers and vortex accelerometers. The Bendix Corporation in particular has for years conducted a wide spectrum of research in vortex dynamics.

In general, however, centripetal technology is unrecognized in today's technology, and the Bendix applications are limited. Centripetal action is characterized by an absence of friction and an increase in energy by the square of its velocity. Unlike centrifugal activity- centripetal forms have a cooling effect and generate diamagnetism or life energy. This is the opposite of magnetic centrifugal energy. Further, diamagnetism is life-giving and constructive not life destroying and destructive.

Viktor Schauberger (see below) has been the primary architect of this technology, which has yet to be adopted even while more powerful and less destructive than heat pressure centrifugal technology. There is considerable evidence that Schauberger technology was sequestered after World War Two. (8) (See Callum Coats' *Living Energies*, which has recently been revived by researchers in Sweden and Denmark as well as Baumgardner in the United States.)

Sound, Vibration, and Structure

The link between sound and matter goes back to ancient times. The Bible, in John 1:1, opens with, "In the beginning was the word," and there is a wealth of wisdom and knowledge behind this phrase. "Word" can be read as "sound" or vibration and vibration is the creative form endowing element in the New Science.

In the late 18th century, German scientist Hans Chladni explored the patterns created by sound on steel plates coated with dust. Chladni found that each sound generated a unique pattern, always the same pattern for each sound.

This creative ordering power of sound is recognized in Chladni and Goethe as natures form endowing element which becomes the sound or chemical ether in Rudolf Steiner and Guenther Wachsmuth. This creative force, totally ignored by modem science is well described by Ernst Lehrs in *Man or Matter* (9):

"[Vibration is] that which is called forth through the etheric forces in nature [and] comprises more than the eternally bounded shape or an organic or inorganic entity. The reason why...in the case of Chladni figures the influence of sound [vibration] causes nothing beyond the ordering of form in our space is because on this plane of nature the only changes that can occur are change in the position of physical bodies. Where the forces of sound in ether form are able to take hold of matter from within they can produce changes of form of a quite different kind. This effect of the activity of sound ether has given its other name, the chemical ether."

Chladni's work was expanded by Margaret Watts Hughes in the late 19th century using vocal sounds, each vocal sound has a specific pattern, some of extraordinary beauty. (10)

Later in the 1960's, Dr. Hans Jenny in Switzerland, a medical doctor, developed an apparatus to vibrate sound through various media and capture the patterns on film. The only variables used were pitch and loudness in various materials.

This extraordinary work on the pattern creating ability of vibration was published as *Cymatics; Wave Phenomena, Vibrational Effects, Harmonic Oscillations with their Structure, and Kinetics and Dynamics.* Initially the two volumes were published only in German (1967 and 1974) and were almost unknown in the English speaking world. In 1985 Brasilus Press and the Institute of Wave Phenomena in Basle, Switzerland produced an English version. (11)

Even then the extraordinary research which has its origins in John 1:1 did not penetrate academic circles in the United States.

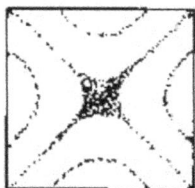

Vibration Modes in a Flat Plate

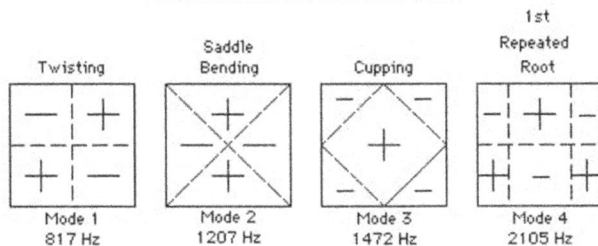

Twisting	Saddle Bending	Cupping	1st Repeated Root
Mode 1 817 Hz	Mode 2 1207 Hz	Mode 3 1472 Hz	Mode 4 2105 Hz

Pattern discovered by Hans Chladni in 1785 (as reported in Ernst Lehr's *Man or Matter*, p. 396). More advanced patterns in Dale Pond's, *The Physics of Love*, and *The Ultimate Universal Laws* (Message Company, Santa Fe, New Mexico, 1996). This is a self-financed work—no government grants or university support.

These books are rooted in the original work of John Keely in the late 19th century which had almost disappeared into oblivion, and which in turn reflected the work of Ernst Chldani and Helmholtz in 19th century German physics.

Dale Pond not only revived this work but explains, so far as one can, the theoretical aspects of sympathetic vibration and how Keely is an early precursor of many modem concepts. These are not books to be read, they are books that need constant and repetitive study. Enlightenment does not come easily.

If your background is music or acoustics you will find a treasure of ideas and practical but unknown research recorded in these volumes.

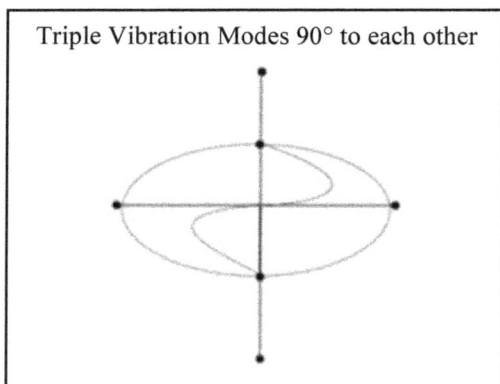

Triple Vibration Modes 90° to each other

Notes

(1) Harry Oldfield and Roger Coghill, *The Dark Side of the Brain* (Element, Dorset 1988 p 126)

(2) Hans Lewy *Chaldean Oracles and Theurgy* (Etudes Augustimenner, Paris 1978) pp. 249-252

(3) Peter Kelder *Fountain of Youth*, p. 11

(4) See Diagram p. 6-3A

(5) Paramhamsa Tewari, *The Substantial Space and Void of Nature* (Harbor Gig Harbor, WA 1989); Of Elementary Material Particles, (Satyasaibaba Publishers, Bombay 1974)

(6) Bruce De Palma has placed much of his work into the public domain.

(7) Details from *The Adams Pulsed Electric Motor Generation Manual*, PO Box 30, Mapleton, Queensland 4560 Australia, U.S. $30.00

(8) Callum Coats *Living Energies* (An exposition of concepts related to the theories of Viktor Schauberger (Gateway, Bath, 1996)

(9) Ernst *Man or Matter* p.397 (Rudolf Steiner Press, London, 1958)

(10) Margaret Watts Hughes *The Eidophone Voice Figures* (Reprint available from Borderland Science Foundation)

(11) Obtainable from Jeff Volk, Lumina Productions, 219 Grant Road, Newmarket, N.H.

(12) *Future Technology Intelligence Report*, November 1996 and issues for October

(13) Videos on sound healing can be obtained from Jeff Volk (above) and Borderlands Foundation.

(14) Richard Gerber, M.D. *Vibrational Medicine*, (Bear& Company Santa Fe, N.M. 87504-2860) 1988

(15) See appendix B for information on Dale Pond.

Chapter 7
Water: Its Memory and Power

A notably significant—yet unknown—aspect of New Science is plain everyday water, its physical properties and relationship with subtle energies. The significance derives from the experimental observation in several countries that water has a memory and this memory is seen to be associated with subtle energies rather than EM energy. Water has quantum properties that make it independent of space and time. These properties are characteristic of homeopathic action, refuting Avogadro's Limit and give considerable credibility to much maligned homeopathy.

The most clearly written short introduction to these quantum properties is an unknown article by Glen Reign "The biological significance of crystal structured water." Rein proposes that: "properties of subtle energy more closely resemble quantum fields than conventional EM fields" and proposes that quantum fields arc involved in transferring subtle energies from crystals to water or biological systems. This would include humans (comprised mostly of water.)

Significantly Rein found with Rama spectroscopy that quantum fields can structure water with specific frequencies which persist over several weeks, and have biological effects.

A similar line of research was undertaken by Patrick Flanagan, Vortex Industries in Flagstaff Arizona based on the original work of Henri Coanda in the 1930s and 1940s. The structured effect is reduced surface tension (i.e., wetter water) and related to micro clusters and their active content of colloidal minerals (similar to the minerals found in Hunza water).

While conventional science accepts structured water and its physical characteristics they have not explored the consequences and remain locked in a prison of their own making. This is because nowhere is the energy nature of water more apparent than in homeopathy and its defiance of conventional Avogadro's Limit.

Homeopathy is based on the principle that "like cures like" i.e., by inducing the symptoms the body generates its own cure. This principle was known to the ancient Greeks but developed into a body of knowledge by a German physician, Samuel Hahnemann (1755-1843).

The effectiveness of homeopathy is derived from the seeming impossible assumption that effectiveness of potency increases with dilution. This is impossible with conventional chemistry because past Avogadro's Limit there cannot remain a single molecule of the material substance in the solution.

One of the principles of modem pharmacology is that there must be significant substance to obtain a therapeutic effect.

The most likely explanation for the homeopathic process assumes that as the physical content is reduced by dilution the 4-space subtle (vibrational energy) increases. For this explanation it is necessary for water to have a memory in order to have subtle energy characteristics.

It is this assertion that extreme dilute solutions are effective that is difficult for modem chemists, pharmacists and doctors to accept. They see it as magic not science, because efficacy is based on the healing action of a chemical compound and for this the compound must be physically present. It is therefore difficult for physical sciences to accept that homeopathic healing is based not on the presence of the material substance but on its active spirit form, its shadow etheric counterpart.

In only one case does conventional wisdom hold good. In synthetic substances the potency curves are straight lines. The conventional effect is noted in synthetic substances only so long as some trace of the material substance is still present. Synthetic substances have no etheric counterpart.

An excellent description of this "primality of spirit" is in Rudolf Hauschka (1) who cites the work of Baron von Herzeele in the 19th century. Herzeele analyzed the increase in mineral compounds in plants grown in distilled water with air bells to keep out contaminated air, i.e., biological transmutation. This book is now extremely rare, only one copy is recorded.

Like Michael Faraday and von Goethe, Samuel Hahnemann was able to visualize the spiritual forces behind the material process, i.e., 4th dimension etheric forces. As early as 1806, Hahnemann understood that a higher dimensional force was released from physical substance as dilution progresses. These forces were no longer tied to the material and indeed Hahnemann referred to them as "almost spiritual" or "spirit like forces".

To quote Hahnemann:

"This mechanical treatment by higher and higher dynamization...brings about final total sublimation into spirit like medicinal which in their crude state can be regarded only as matter—in some cases even as non-medical" (1) *and late "Obviously the measurable sense-perceptible remedial substance disappears to an ever greater extent as potentization proceeds. This means that*

observation focused exclusively on matter one sees only a stepwise dilution in progress, one which permeates all of nature: every natural happening is accompanied by a complementing visible or invisible process. For example a diluting of earthly substances (comparable to a withdrawal from the terrestrial) is balanced by a condensation process of etheric universal processes running parallel to it (in other words the involving of a cosmic element). The forces inhering in plants, minerals and so on ray into a potentizing medium such as a water and bond themselves to it."

The Structure of Water

Water has different structures. These are accepted in conventional science. There is however a structure with liquid crystals (micro clusters) that occurs naturally in water with protein molecules. Glen Rein considers that homeopathic solutions have the properties of quantum fields, not electromagnetic fields. This is more fundamental energy, EM fields being actualized quantum potentials.

These quantum fields have different properties. They are independent of time and space and Rein considers the quantum fields as involved in transferring subtle properties. Rein also argues that water can be charged with specific frequencies and the frequency can be transferred from the water to a biological system.

On November 5, 1985, two French researchers in Paris, Elizabeth Davenas and Francis Beauvais observed an unexpected phenomenon...the second curve. (2) In exploring the biological effects of homeopathic solutions they observed maximum potency at the third dilution (see chart) and then a decrease in potency after this down to the ninth dilution.

However after the ninth dilution the situation changes— the solutions potency re-appears and starts to increase. This was dubbed "the second curve". Between 1985 and 1990 a further 250 experiments were made and "taking everything into account the results presented impressive evidence in favor of the memory of water".

This work was publicized in a book *The Memory of Water* by Michel Schiff, a powerful book that castigates science for censorship, for destroying results not consistent with conventional dogma, the dirty tricks used by the medical establishments in France and England to conceal the unwelcome fact: that homeopathic dilute solutions increase efficacy beyond Avogadro's Limit.

This is illustrated in the diagram from page 26 of Schiff:

THE MEMORY OF WATER

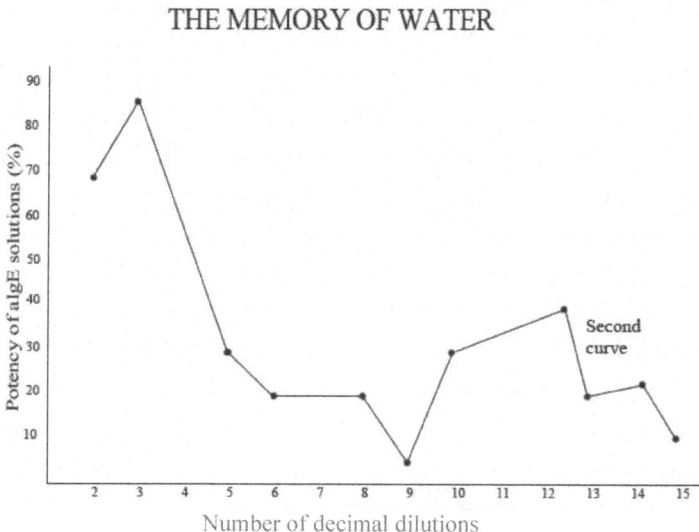

Number of decimal dilutions

After this path breaking experiment was announced, orthodox literature set upon them, squashed or ridiculed the evidence and so Benveniste was largely ignored. (In fact, Harper Collins did not distribute the Schiff book in the United States. FTIR arranged for PACE in Canada and Borderlands Foundation in the U.S. to stock copies). Dr. Benveniste continued this work and made even more extraordinary findings, i.e., that biological activity can be inscribed in water by electromagnetic signal.

In 1997, these are the findings:

(1) Biological activity can be inscribed in water by successive dilutions to install a signal in the molecule.

(2) This signal can be erased with a magnetic field, therefore, it is electromagnetic. The signal is in the kilohertz range.

(3) In hundreds of experiments, the activity of some 30 substances has been transferred to water.

(4) This transfer can be recorded and transferred to another computer via telephone.

What this concludes is that characteristics of matter includes waveforms in the kilohertz range...this has significant implications for medicine, environment and materials science. This parallels research by Harry Oldfield in England who has recorded signals from megaliths at ancient sites. (3)

In *Needles of Stone Revisited* (4) the author Tome Graves considers homeopathy in the light of dowsing and radionics, pointing to the similarity with the radionics work of De La Warr.

As we write in 1997, work at the Benveniste Laboratory is limited by capital and research resources, so the common ground with radionics is probably not being currently explored.

In the United States the work of the Flanagans into the properties of micro clusters and their ability to lower the surface tension of water is not widely known but has merit.

Viktor Schauberger and Living Water (5)

Undoubtedly this extraordinary Austrian naturalist is one of the most under-rated researchers of all time. Schauberger was not an academic, had no degrees. He was a simple forester with an intuitive understanding of the energy in water.

Our present thermodynamic cycle is one of heat-pressure-explosion- and centrifugal behavior. Schauberger's studies of nature led him to understand another cycle, that of cold-suction-implosion and centripetal behavior...far more powerful and life preserving.

The heat pressure cycle is destroying our earth and Schauberger was never able to reach the "establishment" with his warnings.

The alternative suction-implosion cycle is constructive and generates diamagnetism or life energy.

Living Water by Olof Alexandersson (6) has some well written chapters on the difference between our destructive-industrial-imperialist society and the Schauberger constructive-culture.

Among Schauberger's findings are the properties of the spiral. Spiral pipes lose friction. And Schauberger designed the first flying saucer craft, built in Nazi Germany in World War II.

Fortunately, as we enter the 21st century, Schauberger has been discovered There are now numerous books and information. What is needed is a well-financed research program to systematize Schauberger's work. (More in the following chapter). (7)

Schauberger died in the 1950s but his work survived and was effectively developed into devices by Grander, another Austrian naturalist. Essentially Grander has developed effective devices, really radionic filters, which convert bad water into healthy living water simply by changing the information carried in the water.

Notes

(1) Theodor Schenk, *The Basics of Potentization Research*, Mercury Press, New York, 1988, p. 19.

(2) Michel Schiff, *The Memory of Water*, Thorsons, London, 1996.

(3) Harry Oldfield, in *Geosophy*. Video produced by Magic Fire Video, New York.

(4) *Needles of Stone Revisited*, by Tom Graves (Gothic, 1996).

(5) See Callum Coats' *Living Energies* (An exposition of concepts related to the theories of Viktor Schauberger), Gateway, Bath, U.K., 1996.

(6) Olof Alexandersson, *Living Water*, Gateway, Bath, U.K., 1976.

(7) Hans Kronberger & Siegbert Lattacher, *On the Track of Water's Secret* (Uranus Verlagsges, m.b.H., Mossbachergasse 29, A-1140, Wien, Austria).

Chapter 8

Diamagnetism and the Life Force

Diamagnetism is the opposite of magnetism. Little is known about the force because it is not useful for the electromagnetic-heat-pressure-centrifugal-explosion 3-space technologies we use today. It cannot work with heat technologies.

By contrast diamagnetism is prominent in 4-space technology as part of a different thermodynamic cycle, i.e., a cold-suction-centripetal-implosion cycle. This can be visualized as 4-space energy gated into 3-space as a creative, constructive propulsion force.

What is Diamagnetism?

The pole of a magnet exerts two forces: magnetism which attracts substances in greater or lesser degree and diamagnetism which repels substances in greater or lesser degree. In materials science, metals and minerals are classified according to magnetic susceptibility and divided into three groups: nonmagnetic, paramagnetic and diamagnetic. Iron and iron ores are paramagnetic and have the highest magnetic susceptibility. Bismuth and graphite have the highest diamagnetic values and the lowest magnetic values. (1)

Michael Faraday classified materials according to the way in which they were attracted to or repelled by the poles of magnet.

A bismuth rod, highly diamagnetic is suspended by a thread of silk between the poles of an electromagnet. If allowed to swing freely it will position itself at right angles (equatorially) to a line connecting the two poles. (Because the bismuth is repelled by both poles of the magnet).

In contrast, a steel rod suspended in a similar manner between two poles will position itself in an axial position along a line joining the two poles of the magnet.

While known in ancient times, the discovery of diamagnetism is accredited to Sebald Brugmans in 1778, (2) who noted that bismuth and antimony were repelled by the single pole of a magnet.

Michael Faraday made the first extensive explorations and his *Diaries* record the results. (3) Faraday determined that blood, water, and most plant life are all diamagnetic. He concluded there is "a positive and a negative in relation to magnetism".

Twenty years later the Austrian chemist Baron von Reichenbach discoverer of creosote and other chemicals began to explore life force (termed odic) and noted that it was not magnetic.

"To that which supports iron and constitutes the compass let us leave the old name with the original conception of a supporter of iron which belongs to it. If then the term Od should be found acceptable in general use, for the force that does not support iron, and for which we require and seek a name may be easily formed by composition..." (4)

Reichenbach comments on Faraday and diamagnetism:

"There is evidently a tolerable probability that the definition of it I have laid down at ¶215 includes that which Dr. Faraday a year later introduced to the

scientific public as a new material force, under the name of diamagnetism. The British physicist was doubtless aware of my researches, which have appeared in an English translation in London or he probably would have ignored them. Under the definition of the word Od I have comprehended the final cause of all the phenomena described by me, in so far as they have been found irreconcilable with our previous knowledge of the nature of the magnet and the other dynamics and are in particular capable of being extended from magnetic bodies to the so called non-magnetic, to metals, salts and the rest. Diamagnetism was indeed discovered and made known twenty years ago by Seeback, Munke, Buchner, and Becquerel which was also unknown to Dr. Faraday and I have not met in my labors with the transverse position of freely moving nonmagnetic bodies in regard to the magnetic current and for the rest a gap remains between the subjects of our labors. At the same time it is in my opinion not impossible that we are both journeying towards the same point only by different roads." (5)

Von Reichenbach was unwilling to draw any conclusion in 1845 about the nature of diamagnetism, remarking that "these remain questions for further research" and "are deferred till the acquirement of a deeper insight".

In the early 20th century A. Gamgee studied the properties of blood and concluded its constituent compounds were diamagnetic. Numerous studies followed in the 1920-1940 period and recorded by Professor S.W. Tromp in *Physical Physics*. (6)

By the mid-20th century the diamagnetic nature of the life force was known by ignored but conventional biology which based life on chemical compounds and reactions. The 3-space paradigm accepts magnetism but rejects diamagnetism.

Only in the work of Viktor Schauberger in the 1940-1950s do we find recognition of ideas and technology based on diamagnetism. The Schauberger implosion vortex turbine is based on diamagnetism. While Schauberger did not formulate the cold-suction-implosive-centripetal cycle, his work is based on its elements. Moreover Schauberger did understand that the heat-pressure-explosive-centrifugal cycle is destructive while the cold-suction-implosive-centripetal cycle is based on life energy and diamagnetism and is the constructive cycle. (7)

The Usefulness of Diamagnetism

Diamagnetism has not been utilized in technology although a diamagnetic vacuum created by centripetal spiral can generate forces equal to atomic power. The manner in which the forces are created has great significance.

In centrifugal motion, forces are scattered outwards to give rise to friction, pressure and a rise in temperature. This type of motion which generates heat and pressure also generates anti-life force, is wasteful and destructive. It is this outward destructive motion that we have based our technology upon and continued expansion using this cycle leads to the likelihood of even more destructive forces.

The available alternative is impansive, implosive, suction technology, actually more powerful than explosive pressure technology as witnessed by typhoon events. Pressure brings resistance and friction while suction generates no resistance to friction. Suction cools down to the point of anomaly (+4° C) and the contracting spiral generate diamagnetism, the life force. Therefore this alternative thermodynamic cycle, based on diamagnetism is creative rather than destructive, although typhoon phenomena might suggest the opposite.

After long acute observation of nature Schauberger who was a simple forester designed egg shaped vessels or vortex implosion chambers to utilize the centripetal spiral motion in energy creation. The energy is diamagnetic.

The perfect vortex for diamagnetic energy generation is a cross section through the hyperboloid developed from the square hyperbole, i.e., a cross section cut through the throat of a vortex. The ideal spiral is the Kepler harmonic spiral. (8)

Schauberger's contributions show that when a medium, such I as air or water is coiled radio axially (outside inwards) the centripetal contraction generates a temperature drop and diamagnetism. This reaction proceeds in stages. Centripetal contraction first decreases volume and produces condensation. This suction generates a biological vacuum then diamagnetism which can generate an implosive force of major power. This force is not used in today's technology (except possible in some Keely devices). (9) It is found in nature in various types of trees which store this vital energy and release access.

Understanding and use of this diamagnetic force appears to have been concentrated in Austria, first with von Reichenbach then with Schauberger and then with Professors Warburg, Domagk and Ehrenhaft. Today with Grander (10) although interrupted during I the WWII Austrian occupation. It is more widely known as "prana" or Vital Force and in many discoveries as life energy.

An extraordinary feature of our contemporary technology is the manner in which this life force is discovered and rediscovered but ignored in biology. White and Krippner in *Future Science* included a list of almost 100 discoveries of the life force which we expanded to almost 200 discoveries. Because they are not recorded in the biological literature, these discoveries are forgotten, only to be repeated at a later date.

Our biology ignores life energy, we have a dead biology based on absence of life. The hostile reaction by scientists to any understanding of the life force (i.e., N-rays, Reichian orgone, Reichenbach's odic forces or even Sheldrake's morphogenetic fields) is because life energy is the opposing force to the explosive, centrifugal, heat pressure cycle of modem society. Magnetism and diamagnetism are representatives of the ever present duality in nature.

In breathing the chest cavity creates a vacuum which draws air into the body, the air contains the prana or life giving component. The human breathing mechanism parallels the Schauberger diamagnetic spiral turbine.

In the Tibetan rites of rejuvenation speeding up the vortices is one of the steps. The individual spins in a clockwise direction once a day. Too much spinning is destructive, the optimum appears to be a dozen spins. The seven vortices of the body are accelerated by spinning thus restoring the natural slowdown from aging.

However it is in the area of propulsion systems that Schauberger ideas come into their own. The idea of a diamagnetic motor has great advantages. Our present explosion motors use pressure and explosive forces, they generate heat and friction.

A diamagnetic motor would be based on implosion and therefore runs increasingly cool, with no friction and a natural tendency to levitate.

This type of motor coils water into a spiral system of tubes. There is neither radiation nor pollution, and is more suited to domestic standalone units than central power stations.

In fact, these techniques were applied to disc vehicles (flying saucers) in World War II and Schauberger actually built several versions in Austria and Czechoslovakia. Later the Miethe group built several with notable resemblance to the contemporary flying saucer reports.

Diamagnetic saucer propulsion units have remarkable advantages over the explosive, heat pressure units currently used.

The use of suction forces is more efficient, simpler and more effective. There are technical limits to heat pressure technologies. In brief, diamagnetism is cheaper and more effective than rocket and nuclear forces. It is also constructive rather than destructive.

Frictionless Pipes in Schauberger Turbine

(Based on Olof Alexandersson *Living Water* op. Cit page 119.)

Graph of the tests in Stuttgart. The upper curve shows friction in a straight glass pipe, the middle curve, in a straight copper pipe, and the bottom curve, the spiral-wound copper pipe. The solid lines show measured values, and the broken lines the estimated values.

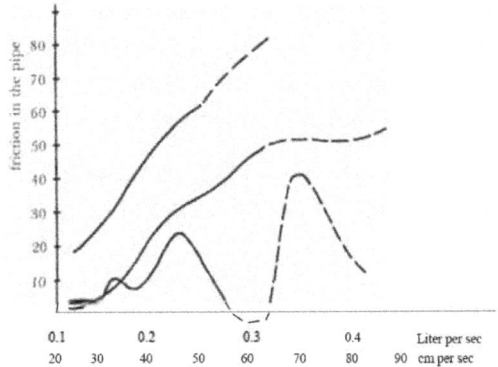

Pipes for liquids and gas. The pipe was also to be used in the "Trout turbine." Its cross section shown in (A), (B), (C), and (D) are different designs of spiral pipe systems showing the pipe to be wound around cylindrical and conical objects. (From Austrian patents no. 19 56 80)

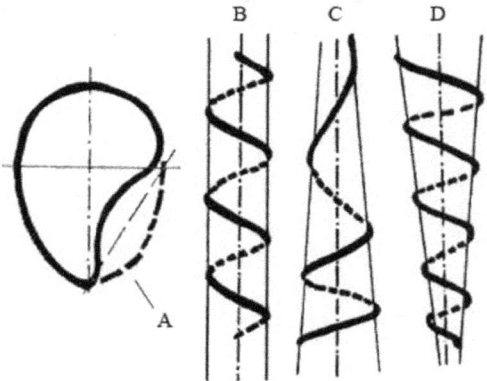

The upper diagram represents tests undertaken at Stuttgart Technical Institute and shows friction in various forms of pipe. Friction is almost absent in a spiral wound copper pipe at a flow of 90 cm per second.

The lower diagram records different designs of spiral pipe used in the Schauberger Trout turbine. Taken from Austrian patent no. 19 66 80.

The Schauberger Turbine

Schauberger ideas in World War II Germany were considered insane and at one point the engineering-architectural establishment demanded he be incarcerated. He was indeed taken to a mental hospital but pronounced sane and given facilities to work on flying saucers powered by a "trout turbine" (and built under command of the S.S.) Briefly, when air or water is rotated into a twisting oscillation, a buildup of energy results leading to levitation. Ideal for flying saucer design power units. Diamagnetic energy has a strong levitation effect and must have received strong research attention within various Governments. Certainly the design of U.S. flying saucers photographed in Nevada and elsewhere suggests suction diamagnetic power rather than "space aliens" or conventional heat-pressure units.

Apparently, some units were built and test flown in wartime Germany. At the end of the war, according to Schauberger, US military units appeared who seemed to understand what was happening and seized everything. (Presumably these were FIAT and BIOS units). Viktor and Walter Schauberger were taken to Texas, along with their models and equipment. There began a disastrous story recorded elsewhere. (11)

Almost unknown, a considerable research effort has been undertaken on Schauberger biotechnical principles outside the United States (i.e., in Austria, German, Denmark and

Switzerland). There is a limited effort inside the U.S. (see Olfo Alexandersson *Living Water*) (Viktor Schauberger and the secrets of natural energy), pp. 148-9 for a detailed listing of U.S. work.

The extraordinary feature of Schauberger work is its constructive life giving nature and its sheer simplicity...shape and water only.

Models of World War II German flying saucers using Schauberger trout turbines. This technology was brought to the United States in 1945 and sequestered. (Source: Olof Alexanderrson *Living Water* (Gateway))

The "Shriever-Hbernmoli" flying disc developed between 1943 and 1945. In 1914, climbing vertically, it reached a height of 12 km in 3 1/2 minutes and a horizontal flying speed of 2000 km/h.

The first test-model developed between 1941 and 1942. This had the same flight properties as that in fig. one, but something was wrong with the controls.

The 'Ballenzo-Schmeyer-Mrethe Disc.' The retractable undercarriage legs terminated in inflatable rubber cushions, it carried a crew of three.

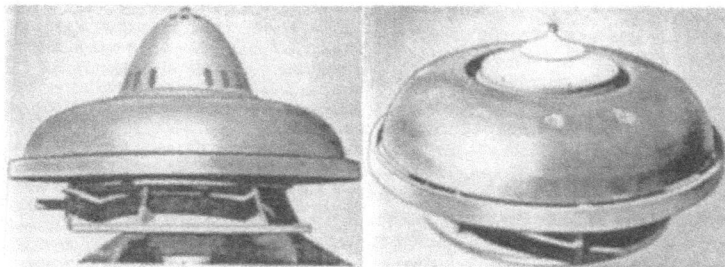

Schauberger's models of 'flying saucers'

Notes

(1) David R. Line, *Handbook of Chemistry and Physics*, 72nd Edition, CRC Press, Ann Arbor, MI, pp. 0-38 to 0-04. Diamagnetic susceptibilities of organic compounds followed by a list of 387 references of research on diamagnetism, mostly from India.

(2) Paul Fleury Mottelay, *Bibliographical History of Electricity and Magnetism*, London, Griffin, 1922.

(3) *Faraday's Diaries*, 1820-1862, Vol. 4, November 1945.

(4) Baron Charles von Reichenbach, *The Dynamics of Magnetism, Electricity, Heat, Light, Crystallization and Chemism in Their Relations to Vital Force*, (New York, Redfield, 1851).

(5) Ibid.

(6) S.W. Tromp, *Psychical Physics* (Elsevier, Amsterdam, 1949).

(7) Callum Coats, *Living Energies* (Gateway, Bath, 1996); Olof Alexandersson, *Living Water* (Gateway, Bath, 1990).

(8) *Implosion Instead of Explosion*, Leopold Brandstatter. Reprinted in *A History of Free Energy Discoveries*, by Peter A. Lindemann, Borderlands, Bayside, 1986.

(9) See Dale Pond in Appendix B.

(10) See Alexandersson above.

(11) Renato Vesco and David Hatcher Childress, *Man Made U.F.O.s*, 1984-1994 (AUP Stelle, 1994).

Chapter 9

Life Energy, the Ether, and Weather Engineering

The most advanced etheric technologies, weather engineering and cloud busting, use somewhat different methods and assumptions but are similar in principle. The basic idea is to utilize – flowing global currents of etheric energy to control weather. These currents are north-south and east-west and vary by time of day and month. (You won't find this in any text book, so don't bother to check).

For some reason this technology raises the most hostility and the most denial. Rejection by Government and academic science in the face of world drought disasters is a sad commentary on a sick world. The State of Malaysia is the only nation that has officially used these techniques and only Greece, Israel, Namibia, Eritrea, Cyprus and Singapore have officially acknowledged operations within their country. Not a single Western world industrialized country has even recognized the technique. This author once discussed the possibility with South African Government officials but the project was not pursued.

This technology can not only induce rainfall (to control drought and forest fires) and limit rainfall but also control smog (at minimal cost) and floods. There is a probability that cyclones and hurricanes can be at least partially controlled

There are several primary energy weather modification technologies and while essentially similar the two major operators prefer to use their own assumptions and descriptive terminology.

Dr. James DeMeo based in Ashland, Oregon, and founder of the Orgone Biophysical Laboratory, uses the classic Reich

94

cloud buster technique for reversal of desertification by encouraging adjustment to normal weather patterns.

Trevor James Constable has introduced electronic controls and Steiner etheric ideas and married them to Reichian ideas on biological energy, and has now extended seaborne and land operations to successful air trials.

Both these leaders in the field has a group of dedicated associates here and abroad and have operated successfully if quietly, around the world for many years.

Their results are extraordinary and make cloud seeding, a polluting chemical method of questionable reliability, look like the horse and buggy era.

The original discoverer was the persecuted genius Dr. Wilhelm Reich. Rarely is anyone attacked by ALL political factions as was Wilhelm Reich. He was attacked by Communists and Nazis. The Governments of Norway, Denmark and the United States were hostile. He died in a U.S. jail after being railroaded by the Food and Drug Administration. His books and papers burned by court order: This is extreme but it does show the hostility and anger aroused by these new technologies.

Reich was a non-violent man. His only weapons were ideas and words. Yet Reich was seen as a deadly enemy...and the successors to the 1950's FDA which jailed Reich have the same attitude and still rule today's bureaucracies.

His memory is continued in the Reich Museum Rangeley, Maine.

Another outpost of Reich research today is the Greensprings Research and Educational Center, at the Orgone Biophysical Research Laboratory, Ashland, Oregon, founded in 1978 by Dr. James DeMeo. This Center publishes a range of research reports including the Orgone Accumulator Handbook, now translated into Japanese, Greek, Spanish, Portuguese and German and four volumes of *Pulse of the Planet*. Research links to parallel work by scientists as Harold Burr, Giorgio Piccardi, Louis Kervran and others.

One remarkable aspect of DeMeo's work is a series of drought abatement experiments in Greece, Namibia, Eritrea and Cyprus, desert greening experiments in the U.S. and elsewhere and pollution experiments in Germany.

Like other aspects of the new science this established research center owes much to the dedication and perseverance of an individual James DeMeo and a group of supporters in the orgonomy movement, scattered around the world. Quietly over years this group has been working to reverse desertification and gives genuine help where it is needed. Their reward? Harassment and Internet slander: We have long thought that the origin of the harassment was a private corporation. Recently more ominous signs appear. The slander surfaced partly through CSICOP, an organization which has in the past been used by CIA as a front. Unless CSICOP and CIA clean up their act and disavow harassment activities without doubt, in the new world now emerging, these organizations will be seen as destroyers of national security, not protectors.

The other operator is Trevor James Constable, whose operation is weather engineering by manipulating etheric currents. The orgone energy of Reich and DeMeo is the same energy under a different name—the life energy ("orgone" = "chemical ether").

A distinction is made between the DeMeo-Reich method and the Constable etheric method.

DeMeo states that cloud busting is a technique "to restore the lost property of atmospheric pulsation" and restore atmospheric self-regulation. Cloud busting aids the atmosphere to act in a more natural free self-regulating manner. This is foreign to meteorology which starts with the assumption that we live in a dead world, empty of any natural energy

For orgonomy, the world is a living entity and this is the basis of orgononomic weather work. Whether one agrees, or not the fact remains that the weather can be influenced using orgonomy as a theoretical base. We will briefly describe one operation by each operator (DeMeo and Constable) to suggest the advanced state of this technology.

The Work of James DeMeo

The years 1988 to 1991 were abnormal in the Eastern Mediterranean (Greece, Turkey, Egypt and Israel), as these countries experienced extended and severe drought. So severe was this drought that on Cyprus in early 1992 water delivery was restricted to 2 hours PER WEEK. In early 1991 a group of American orgonomists, including Dr. Walter Hoppe, Dr. Richard Blasband, who had previously worked in Israel, got together with Dr. James DeMeo to establish a program that would reinforce the natural tendencies towards rain in late 1992.

A "small but powerful cloud buster" was built in an Israeli machine shop. The weather by November 15 had developed into a hardened desert-like character. In coordination with the Hellenic Orgonomic Association in both Greece and Cyprus, operations began. By November 17 rain began in Greece and by November storms generated in the East Mediterranean moving towards Israel. Rain started to fall on November 27th and by December 7, accumulated rainfall was something like 200%-400% of normal (See map).

On February 15 the news reported: "It's official. The current and far-from-over winter is the wettest in a century in Jerusalem and Tel Aviv, there hasn't been so much rain since records were first kept in 1904."

DeMeo has since repeated this type of operation in Eritrea and Namibia, quietly and without publicity, and at no public expense. The operations are entirely privately financed.

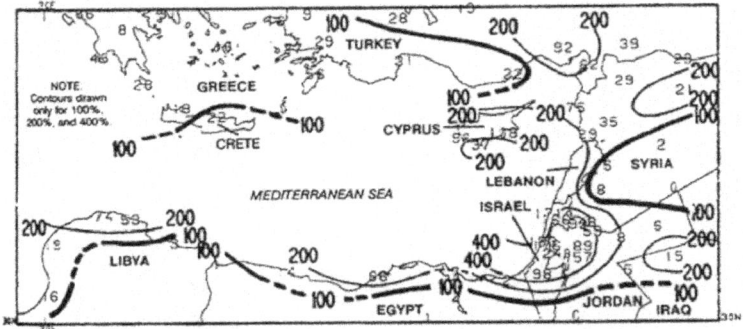

Percentage of Normal Precipitation Map Eastern Mediterranean, Nov. 27 to Dec 7, 1991, showing rainfall contours for the first major storm to enter Israel shortly after the onset of cloud bursting operations. This exceptionally heavy rainfall episode began only 10 days after cloud bursting operations were initiated, and only a few days after those operations ended. The storm which brought these rains had developed in the western and central Mediterranean as a "cut-off low" on approximately Nov. 21-22, during the cloud bursting operations. It slowly moved eastward, intensifying as it approached the cost of Israel. Six additional episodes of moderate-to-heavy rainfall entered Israel after this storm, from mid-December through March of 1992, and these are identified on the graph below.

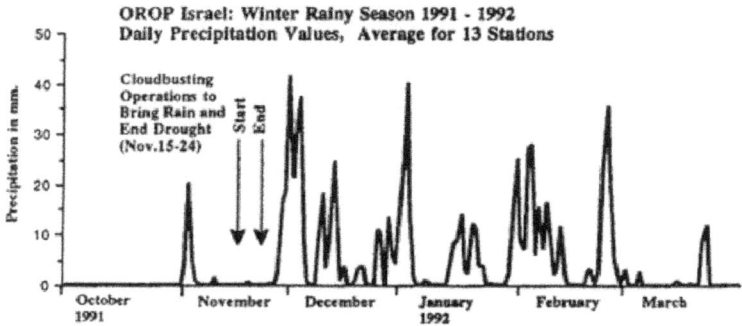

Daily Precipitation Values, Average for 13 Stations in Israel, for the period from 1 October 1991 to 31 March 1992. The periods of cloud bursting operations are marked with arrows and roughly six major pulses of rainfall can be identified on the graph. Only the first of these rainfall episodes, that of Nov. 27 to Dec. 7, is plotted on the map above.

99

The Work of Trevor James Constable

Trevor James Constable came through the Reich school of orgonomy and was for a time its Director of Research. Constable moved away from pure orgonomy to incorporate Steiner and Wachsmuth anthroposophic principles coupled with various electronic devices. Constable technique is engineered with far more reliability than cloud seeding and chemical assault. Both DeMeo and Constable methods are non-polluting.

Etheric weather engineering uses bio geometric translators, simple juxtaposed geometric shapes based on Golden Proportions. The translators dam up the etheric flows and force water discharge. Simple, yet extremely effective.

For over ten years Constable and his group have conducted tests in the Pacific Ocean and Far East. Many tests used the Matson Lines Container ship SS MAUI and many are recorded on video tape. The most dramatic video is an artificial storm generated by Constable translators and which accompanied the ship on its track across the Pacific (the so-called "Moses effect").

While difficult to describe in writing the dramatic effect of etheric procedures is brought to life by radar and video. The Constable volume, *The Loom of the Future*, is crammed with photographs, radar reproductions and other graphic representation of the various programs and operations.

All Constable operations in the U.S. are reported to National Oceanic and Atmospheric Administration on NOAA Form 17- 4, so there is no question that the Federal Government is well aware of the potential for these devices. It is beyond this authors understanding how the Federal Government can allow atmospheric disasters to ruin citizens and yet take no action.

For political and legal reasons Constable has suspended operations in the United States...and we spend something like $400 billion a year on "national security."

Operation PINCER II

This operation is probably the best designed and most effective of all Constable projects.

Statistically the Los Angeles Basin has no rain in July and records exist for 100 years. In brief, July is an excellent control base for any Los Angeles project. If you generate rain against a control of zero then SOMETHING is happening out there. Even a Federal bureaucrat should be able to see that.

Constable designed PINCER II to bring rain into the Los Angeles Basin from subtropical Mexico. In July the normal flow of moisture is northeast from Mexico into Arizona. The plan was to divert Mexican origin moisture into the dry Los Angeles basin. The plan was filed with the National Oceanic and Atmospheric Administration on Form NOAA 17-4 (Reproduced overleaf). The results were extraordinary, as L.A. experienced the most rain for any July in Los Angeles history (see documents attached).

Proof? Not for the Federal Government (although it was recorded in their own radar stations) or the Los Angeles Board of Supervisors. Between the politicians and the bureaucrats the citizen and his wants are forgotten.

Document Annex for Operation PINCERII

(Permission granted by TJC-ATMOS Inc., Singapore

Primary Energy Weather Engineers)

(1) Federal Form NOAA 17-4, dated 20 June 1986.
(Authors note: The U.S. government cannot claim it has no file information.)

(2) Safety program for PINCER II operation.

(3) Planning map for PINCER II

(4) Summary of results (written by Trevor Constable).

TO: Atmospheric Programs Office, RD2 National Oceanic and Atmospheric Administration Rockville, Maryland 20852	NOAA FORM 17-4 (4-81)	U. S. DEPARTMENT OF COMMERCE NAT'L OCEANIC AND ATMOSPHERIC ADM.
1. PROJECT OR ACTIVITY DESIGNATION, IF ANY Operation PINCER II	**INITIAL REPORT ON WEATHER MODIFICATION ACTIVITIES** (P.L. 205, 92ND. CONGRESS)	

3. PURPOSE OF PROJECT OR ACTIVITY : Research, Rain, L.A.Basin, Rain So. Calif. Nat. Forests (Anti-Fire) & Humidity	**2. DATES OF PROJECT**	
	a. DATE FIRST ACTUAL WEATHER MODIFICATION ACTIVITY IS TO BE UNDERTAKEN	1 July 86
4.(a) SPONSOR Elevation, San Diego County,	**b. EXPECTED TERMINATION DATE OF WEATHER MODIFICATION ACTIVITIES**	31 July 86
NAME TJC-ATMOS INC.	**4.(b) OPERATOR** **NAME** SAME	

AFFILIATION Independent Consultants	**PHONE NUMBER** 213 833 4260	**AFFILIATION**	**PHONE NUMBER**
STREET ADDRESS 3726 Bluff Place		**STREET ADDRESS**	
CITY San Pedro	**STATE** Calif. **ZIP CODE** 90731	**CITY** **STATE**	**ZIP CODE**

5. TARGET AND CONTROL AREAS (See Instructions)

TARGET AREA		CONTROL AREA	
LOCATION see attached engineering drawing	**SIZE OF AREA** sq. mi.	**LOCATION** This concept is not applicable to these methods.	**SIZE OF AREA** sq. mi.

6. DESCRIPTION OF WEATHER MODIFICATION APPARATUS, MODIFICATION AGENTS AND THEIR DISPERSAL RATES, THE TECHNIQUES EMPLOYED, ETC. (See Instructions) Further developments of juxtaposed geometric forms for direct insertion into and manipulation of, the primary energy (etheric) continuum that underlies the atmosphere. Apparatus employs no electric power other than that for physical rotation of "Flying H", fixed base units. The latter are sited at 4 points as per attached drawing. Additional mobile units are carried in two "gun cars" for operation while traversing highways. NO CHEMICALS, NO RADIATION employed anywhere in this project.

7. LOG BOOKS: Enter name, affiliation, address, and telephone number of responsible individual from whom log books or other records may be obtained.

NAME TREVOR J. CONSTABLE	
AFFILIATION PRESIDENT, TJC-ATMOS	**PHONE** 213 833 4260
STREET ADDRESS 3726 Bluff Place	
CITY San Pedro	**STATE** CA **ZIP CODE** 90731

THIS REPORT IS REQUIRED BY PUBLIC LAW 92-205; 85 STAT 735; 15 U.S.C. 330b. KNOWING AND WILLFUL VIOLATION OF ANY RULE ADOPTED UNDER THE AUTHORITY OF SECTION 2 OF PUBLIC LAW 92-205 SHALL SUBJECT THE PERSON VIOLATING SUCH RULE TO A FINE OF NOT MORE THAN $10,000, UPON CONVICTION THEREOF.

8. SAFETY AND ENVIRONMENT

☐ YES ☒ NO Has an Environmental Impact Statement, Federal or State been filed? If applicable.

☒ YES ☐ NO Have provisions been made to acquire the latest forecasts ...ional Weather Service, Forest Service, or other ... if yes, please specify on a separate sheet.

☒ YES ☐ NO Have any safety procedures ...or suspension of operations, monitoring methods, etc.) and any envir ...to the possible effects of the operations) been included in the operation ..., please furnish copies or a description of the specific procedures and guidelines.

OFFICAL ADVANCE NOTICE TO NOAA

9. OPTIONAL REMARKS (See Instructions, Use Separate Sheet.)

NAME TREVOR J. CONSTABLE	**CERTIFICATION:**	I certify that the above statements are true, complete and correct to the best of my knowledge and belief.
AFFILIATION TJC-ATMOS INC.		**SIGNATURE** Trevor Constable
STREET ADDRESS 3726 Bluff Place		**OFFICIAL TITLE** President
CITY San Pedro	**STATE** CA **ZIP** 90731	**DATE** 20 June 86 213 833 4260

NOAA FORM 17-4

TJC-ATMOS. INC.

PRIMARY ENERGY WEATHER ENGINEERS AND CONSULTANTS

1805 EAST CHARLESTON BOULEVARD, LAS VEGAS, NEVADA 89104 U.S.A.

A NEVADA CORPORATION

ADRESS ALL CORRESPONDENCE TO:

| 3726 BLUFF PLACE |
| SAN PEDRO, CA |
| 90731 |
| U.S.A |

AMPLIFICATIONS TO INITIAL REPORT ON WEATHER MODIFICATION ACTIVITIES (NOAA FORM 17-4), FOR OPERATION PINCER II

Under Section 5 TARGET AND CONTROL AREAS:

While valid for nucleation and kindred methods, this section is inapplicable to primary energy weather engineering, in as much as an entire region is affected by successful operations. The main control for this operation is the statistical rainlessness of the Los Angeles Basin area in the month of July.

Under Section 8 Safety and Environment:

24-hour alarm surveillance is maintained on NWS VHP radio stations throughout these operations. All bulletins, forecasts, flash flood watches and warnings are closely followed. Radio weather facsimile is used to copy satellite photographs, etc. Close touch is maintained by telephone with NWS radar at Palmdale in the event of severe conditions. In addition, continuous digital readout is available of primary energy continuum flows. This Swiss-developed instrument is unorthodox, but experience has proved that it provides highly useful lead time of forthcoming precipitation.

Safety procedures provide for the immediate total deactivation of all modification units and related activity in the presence of NWS warnings via VHP.

Copies of the Initial Report and associated engineering drawing and these amplification, are being supplied to the NIC at Los Angeles, the MIC at San Diego, and the NWS Radar at Palmdale.

Since the Los Angeles Basin is statistically rainless in the month of July based on more than a century of observations, environmental hazards are considered to be minimal, but every precaution is being taken in the event of a successful modification of the July weather.

104

SUMMARY OF PINCER II

The statistical facts of Pincer II can be found in NOAA's official publication. CLIMATOLOGICAL DATA. CALIFORNIA. July 1986, Volume 90, Number 7 The official stations covering the operational area of Pincer II are listed in two statistical divisions – South Coast Drainage and Southeast Desert Basin virtually all of Southern California is covered in these two Division which contain 100 official stations.

72 OF THESE 100 STATIONS RECORDED RAIN IN JULY 1986

Largest rains were at Anza with 3.20" (80mm) and Cuyamaca with 2.15" (53.6mm)

Other Notable rainfall in the depth of the Southern California dry season was

Idyllwild Fire Station	43mm	1.72" (Adjacent to Ft. Zinderneuf)
Dagget: FAA Airport	29mm	1.18" *
Borrego Desert Park	23mm	.91" *
Henshaw Dam	27mm	1.07"
Deep Canyon	28.2mm	1.13"
Iron Mountain	36mm	1.47" *
Niland	37.5mm	1.50" *
Julian	24.5mm	.98"
Parker Reservoir	27.5mm	1.10" *

* Desert station

The 18" (6.5mm) of rain at Los Angeles Civic Center, the bull's eye in the profiled target zone, made July 1986 the wettest Los Angeles July in 100 years. The statistical control provided by more than a century of official records was invaluable. Pincer II showed, for the first time in history that an etheric weather engineering operation could be:

1. Designed with statistical controls
2. Presented in an engineering drawing
3. Filed with the Federal authorities prior to commencement
4. Executed objectively, exactly as depicted in the pre-filed notification

Official records, including the National Weather Service radar maps presented herewith, conclusively prove that the enormous masses of moisture coming out of Mexico were diverted more than 220 miles away from their normal pathway, which minimal rain followed the normal July "Blue Route" into Arizona.

As Specified in the pre-filed engineering drawing, the rain masses wound up over the Los Angeles Basin. The Federal radar maps verify that the thundery masses were brought up first along the interior "Red Route". The moisture curved into the Los Angeles Basin from the south and southern. This was one era of the PINCER.

Life Energy, the Ether, and Weather Engineering

Summary 2

Radar verifies that later masses of thundershowers followed the pre-specific and highly anomalous "GREEN ROUTE" along the southern California shore, on the ocean side of the mountains. This was the second era of the PINCER. The spectacular PINCER from landward and seaward is unequivocally shown in the NOAA radar maps just as it appears in the pre-filled engineering drawing.

The radar depictions further show how the coastal "GREEN ROUTE" rain masses were manipulated at Point Fermin so that they passed mainly east of the Point and northward up to Los Angeles Civic Center. The distinctive curve in the rain mass at Point Fermin attest to the decisive diverting effect of the Point Fermin installation.

Official records of lightning strikes for 22/23 July 1986 corroboratively reveals that an anomalous. 2 to 1 preponderance of all the lightning strikes in the Los Angeles area that night hit in the immediate vicinity of Point Fermin. A compelling photographic and video record was made of this activity by the Pincer II chief engineer, who knew what to expect.

PINCER II verified the validity and value of etheric weather engineering as a revolutionary new technique in the environmental struggle. The Federal authorities were notified before and after, as required by law, and these reports were dutifully filed by the bureaucracy. PINCER II passed into history, the best rain operation that the Constable group had carried out up to that time in the USA.

Operation PINCER II did not really end with the statistical and esterological victory in Los Angeles and vicinity. Engineering weather is oblivious to political and geographic boundaries. Once set in train by operations like PINCER II, the engineered systems follow their own natural love and lifespan. The rain produced by PINCERII continued east and northeast out of California and duly fell in at least eight states beyond California. Twenty years of such engineering work in Southern California watching "downstress" effects after a regional operation leaves no doubt that primary energy weather engineering is national in scope. A weather engineer must think in continents.

Such wide-space thinking is enjoined by the ever-increasing weather engineering power and steadily distinguished size of etheric weather engineering apparatus. Insurmountable politico-legal problems in the USA arising from the regional character of etheric weather engineering operations let to the abandonment of commercial work in the USA and the pre-filing of rain operations.

The Constable group since 1986 has concentrated its U.S. work primarily on the mastery of smog. Three highly effective southern California smog operations were counted in 1987, 1989 and 1990.

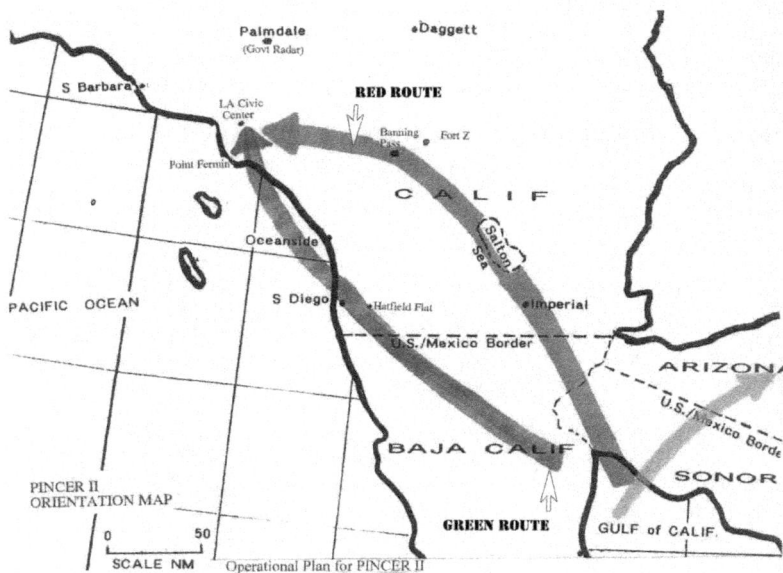

PINCER II
ORIENTATION MAP

Operational Plan for PINCER II

Chapter 10
Etheric Technology and Transducers

The ether is that all pervading universal energy known for centuries and now rediscovered by quantum physicists and the New Science. Forces and attributes within the ether are referred to as etheric, here termed four dimensional or 4-space. The physical realm a counterpart of the etheric is termed three dimensional or 3-space.

Guenther Wachsmuth suggests (1) in the preface to his classic commentary on the ether that the power of knowledge of the higher forms of reality in Nature must be <u>awakened</u> within the individual. The higher reality is not immediately known. Awareness requires attributes which are not the same as those required for recognition of the mechanical or the physical which are readily observable by all individuals. According to Wachsmuth the theory of the etheric is the "Master key to the knowledge of Nature". (2)

Possibly this may explain why orthodox scientists exhibit such emotional distress when presented with 4-space phenomena. Etheric activities are to them impossible and fraudulent <u>even if clearly and undeniably observable.</u> An example is the 1989 announcement of cold fusion by Fleischmann and Pons...most scientific observers did not have the inner awakening necessary for recognition. So they emotionally denied what was observed and termed the discovery "fraudulent", "bad science" and so on.

The phenomena of 3-space are readily observable. Events originating in 4-space require specific talents or sensitivities not immediately available...they have to be developed and nurtured.

What is the ether that originates these 4-space activities?

Before Einstein came along the ether was generally seen by most physicists as an elastic solid in the fourth dimension. Our concept of 4-space is not new, it has merely been buried by the overwhelming and all too visible onslaught of materialist technology.

In the early 20th century Claude Bragdon in *Four-Dimensional Vistas* noted that:

> *"It assists the mind to think of the ether as four dimensional because then a light wave would be a superficial disturbance of the medium - superficial but three dimensional as must needs be the case with the surface of a four dimensional solid".*   (3)

Bragdon also pointed out that the early 19th century mystic Swedenborg announced a Higher Space Theory and quotes Swendenborg:

> *"It is observed that the natural world exists and subsists from the spiritual world, just as an effect exists from its efficient cause".* (4)

Gustav Le Bon in *Evolution of Matter* wrote a two chapter review of this immaterial basis of the universe, a rather vague overview because as Le Bon suggests "due to the fact that this immaterial element cannot be connected to any known thing". (5)

While Dmitri Mendelieff visualized the ether as a gas (6) with an atomic weight one millionth that of hydrogen, Le Bon claimed that "the properties of the ether do not permit it in any way likened to a gas. Gases are very compressible and the ether cannot be so". (7)

However, the late 20th century working experience of Trevor James Constable with etheric weather engineering is that the ether is indeed compressible.

The physicists of the 19th century hardly agreed on <u>any</u> property of the ether. Maxwell supposed little spheres in rapid rotation. Fresnel visualized constant elasticity and variable density.

In any event, it was typical to see the physical phenomena of heat, light, electricity, magnetism and gravity as disturbances in the equilibrium of this ether. The only way we know the ether from physical 3-space is as a disturbance in this equilibrium and 19th century physics saw vortices and rotatory motion as the best way to understand etheric equilibrium. In fact, Le Bon touched on the topic of this present book by suggesting that bicycles and helicopters achieve equilibrium against the force of gravity by rotatory or forward motion. The 19th century investigators also determined experimentally the geometric forms assumed in the translation phase, again confirmed by Constable in the late 20th century and discussed in detail by Wachsmuth.

The common 19th century conclusion was as stated by Le Bon, "the ether is a solid without density or weight however intelligible this may seem".

C. Howard Hinton in *The Fourth Dimension* relates to 4-space to electricity:

"Thus on the hypothesis of a fourth dimension, the rotation of the fluid ether would give the phenomenon of an electric current. We must suppose the ether to be full of movement, for the more we examine into the conditions which prevail...the more we find that an unceasing and perpetual motion reigns".

Hinton concludes as follows:

"If we suppose the ether to be filled with vortices in the shape of four dimensional spheres rotating with an A motion, the B motion would correspond to electricity in the one fluid theory". (8)

Hinton adds that our three dimensional world is "superficial" and the processes "at the basis of the phenomenon of matter only reveal themselves by their motion". (9)

By the late 19th century the significance of rotary motion and the vortex was discussed and etheric manifestation in the physical world seen as a consequence of rotary motion. Gustav Le Bon summed it as: *"it is in rotatory motion that is found the best explanation of the equilibrium of atoms".* (10) William Tiller, Professor Emeritus at Stanford University, picked up these ideas and made considerable theoretical extension in the late 20th century. Originally, Tiller visualized physical and etheric dimensions only. Later he expanded these concepts in a little known article of great significance. (11)

The physical-etheric dimensions were expanded into multiple dimensions with positive and negative space/time frames and varying characteristics and manifestations.

The etheric dimension has today been reintroduced as the quantum vacuum or a less well-known name zero point energy. Modem leading edge physics has rediscovered the classical ether physics based on an immaterial ether. However modem physics has yet to move into exploration of the ether and its formative forces, none of which are explained in orthodoxy at the turn of the 21st century.

An interesting exception to this statement is the quantum computer underdevelopment by ARPA (Advanced Research Projects Agency) in early 1997. Quantum mechanics states that particles have superposition, i.e., in a quantum world they have no fixed position and hold all positions simultaneously. (Not A or B or C but A and B and C). Only when measured do they adopt a single location in 3-space. Theoretically although quantum particles could be used as quibits in design of a computer the superposition condition may make this practically impossible. The bits won't have a time/space attribute. In brief, quantum physics wants to cross the line from 3-space to 4-space using 3-space methods. One wonders if ARPA considered "remote computing", a parallel technology to "remote viewing".

If computer science follows the technological pattern suggested by other areas where the etheric uses intent and rotation then the quibit approach will not work. However this $5 million ARPA study will generate a wealth of information about the ultimate 3-space frontier. It could be the most revealing $5 million project ever commissioned by ARPA.

If superposition precludes construction of a quantum computer, ARPA is not home free in protecting its code breaking techniques. We see the etheric world as a world without secrets so the attempt to construct a quantum computer is ultimately an exercise in futility while simultaneously an expedition into 4-space reality.

In contrast another neglected philosopher, Rudolf Steiner with his secretary Guenther Wachsmuth and a Swiss medical doctor, Ernst Marti outlined a theory in which the etheric is the architect and builder, i.e., the creator of the physical. Today materialist physics is in a blind alley, i.e., all creation is seen as random chemical combination. But a little known philosophic movement has developed clear expression of the ether, its parts, functions and spiritual origin.

Wachsmuth elaborates a body of formative (creative) forces and Marti asserts that the ether does not appear naked but clad in these formative forces.

Our exploration of 4-space technologies as gated into 3-space suggests that Steiner, Wachsmuth and Marti were very much on a line of reasoning that explains what one can observe in 3-space anomalies. One can identify a life force in biology, a telluric force in dowsing, a creative shaping force in vibration physics, an information force in psychic mind experiments, a force with affinity for water in Constable weather engineering. Moreover in low energy nuclear transformation, we can identify the ideas of the old alchemists, close to nature, yet derided by modem science.

A theoretical analysis of the ether based on Rudolf Steiner expanded by Wachsmuth is now the basis of anthroposophic science. (12)

In the Steiner-Wachsmuth analysis, the ether is in four forms, in two groups, with characteristics entirely different from those of matter in solid, liquid or gaseous forms. The ether is not matter— it is pre-matter.

(1)   warmth ether is not motion but a concrete formative force, related to the heat state of aggregation.

(2)   light ether calls forth for the human eye the phenomenon of light and is associated with the gaseous state of matter.

(3)   chemical or sound ether is the formative force in chemical processes, differentiation, dissolution and union of substances. The force transmits through sound or tone and is thus the formative force which creates Chladni figures where frequency is directly related to shape and form. The phenomenon of cold suction is associated with the chemical ether, and also associated with the liquid state of matter. Life

ether, another cold suctional force and according to Wachsmuth linked to gravitation, levitation and to the solid state of matter.

Viewed through the eyes of Steiner's four ethers we find that 3-space physical processes are actually no more than *inefficient imitators* of etheric activity. For example, in 3-space the physical transmutation of elements, or atomic transmutation, can only occur using great heat and pressure according to physics. The ultimate absurdity being the construction of gigantic multi-billion dollar linear accelerators as the only means to explore the inner atom. However, as we go to press, so-called table top transmutation, or low energy nuclear transformation, is emerging (though long known through the work of Louis Kervran). This low energy nuclear transformation has etheric qualities.

In homeopathy, Jacque Benveniste of Paris found experimentally what homeopathists had long known: that dilution increases efficacy. Michel Schiff, in his remarkable 1995 book entitled, *The Memory of Water* (13), demonstrated that after the ninth dilution a "second curve" of efficacy emerges where no molecule of the original physical compound remains. Logically, some other force takes over with the memory of the original compound, i.e., an etheric explanation.

Our technological exploration for physical processes developing an energy or an activity unexplainable by man-made physical laws suggests a variety of gating mechanisms or transducers that enable us to tap 4-space from 3-space. These gates include, but are not limited to, rotation, vortical motion, biological process and the mind itself for some combination of these transducers. Many processes are suggestive of the chemical (sound or number) ether at work.

In fact, Trevor Constable, whose etheric weather engineering is among the most advanced of 4-space technologies, specifically links weather engineering and geometric translators to the chemical or sound ether. What is truly extraordinary is that decades of precise, measurable recorded self-financed work by Constable and his group has gone almost entirely unrecognized. (14)

Our experience has been that the Constable technology is so powerful, so awe-inspiring that for many, especially physical scientists a process, of denial and hostility immediately emerges. Captain David Morehouse, formerly of U.S. Army and Defense Intelligence Agency, a notably successful remote viewer with over 250 missions for the U.S. government, constantly refers to the ether as the medium in which he travels on viewing missions (15). Captain Morehouse was discharged by the US Army after a valiant career because he would not follow man-made classification procedures. The Army was flexible enough to use the technology but not smart enough to understand that the etheric is infinitely more powerful than the physical, and U.S. government demands or laws have no jurisdiction in 4-space.

One can judge how little the US Government understood what it was dealing with when we note that DIA placed the technology into its Psychological Operations Unit where Colonel Michael Aquino, a well-known practitioner of "black arts" and one time involved in messy San Francisco child abuse scandals, is located. A staffing decision that guaranteed failure.

In attempting to keep the technology secret the Defense Intelligence Agency itself guaranteed public access. There are no secrets in 4-space. And Black Arts are destructive whereas the ether has only positive constructive attributes.

Altered states of consciousness where the individual accesses another dimension have used rotation. The ancient Chaldeans used rotating tops (gyroscopes) for prediction. The Sufi whirling dervishes use body rotation to enter a trance state. Tibetan longevity exercises use rotation. (16)

Gyroscopes and gyro stabilizers use rotation to achieve 4-space properties that cannot be achieved in 3-space. The notably closed nature of the scientific mind is demonstrated in the fact that of tens of thousands of scientists who played with spinning tops as children, only a handful investigated the obviously non 3-space properties: the ability to defy gravity when precessing. Professor Laithwaite at University of London, Bruce De Palma, one time at MIT, are the only investigators that we know of who have been intrigued by 4-space gyroscopic and levitation phenomena.

Another dramatic example we describe above is the N-machine. Over-unity electricity is generated by rotation of magnetic cylinders. Over-unity is not consistent with the Laws of Thermodynamics in the physical dimension. Again, while originated by Michael Faraday it was developed by Bruce De Palma, Paramhamsa Tewari in India and Shiuji Inomata in Japan and ignored by 3-space orthodoxy. We can understand this in terms of Steiner, that our physical phenomena are merely inefficient duplicates of the etheric, an insight reflected in William Tillers concept of duality in physical forces.

Finally, in our brief description of Viktor Schauberger, we locate another thermodynamic cycle. Our 3-space thermodynamic cycle generates technologies based on heat, pressure, explosion centrifugal activities. Schauberger identified another cycle reflecting cold, implosive centripetal, suction activities. We found no record that Schauberger was aware of Steiner or Wachsmuth, yet the

Schauberger cold suction cycle is descriptive of the chemical ether, while our 3-space thermodynamic cycle is reflective of warmth ether tendencies. Frankly there is an area here we have not explored but intuitively we sense a link between Steiner and Schauberger.

When we place our transducer gates in a diagram, suppose in Diagram 10-1 we identify various anomalies which do not conform to 3-space laws and insert the gate or transducer type between the 3- and 4-space areas. Somewhere in this boundary region are the Akimov torsion fields where objects and events in the observed 3-space world interact with the quantum 4-space world. This torsion field is presumably the region of enmatterisation or pre-matter. It is in this boundary area where we find leading edge quantum mechanics and particularly quantum computers. Because of superposition the bits in a quantum computer (if one could be built) are not 0 or 1 but 0 and 1 at the same time.

In the physical world a bit can only be one or the other. In the etheric world without time and space the bit becomes a quibit and can have any location (or superposition) or all locations simultaneously. In other words, quantum physics is approaching the 4-space realm from 3-space and trying to develop the quantum computer may tell us more about the area between the dimensions.

DIMENSIONS, TRANSDUCER GATES & RELATED J-SPACE TECHNOLOGIES

		MIND	MIND	DIAMAG-NETISM	ROTATION	ROTATION	ROTATION	GIXMETRIC	SUCCUSSION WATER
1	4-SPACE ETHERIC DIMENSION	ETHER (PRE-MATTER)							
2	THE TORSION FIELD AND TRANSDUCER	VORTICIES IN THE TORSION FIELD (ARTMON)							
3	GATES	PHYSIC	REMOTE VIEWING	HOLOCK TURBINES	LIFE ENERGY	GYRO	N-MACHINE	LONGEVITY	WEATHER ENGINEERING
4	3-SPACE PHYSICAL DIMENSION	ALTERED STATES PRE-COGNI-TION TELE-PATHY	U.S. ARMY DIA, CIA SWANY	SCHAUM MERGER BAUM-GARDNER	VON REICHEN RACH REICH DEMEO	GYROSCOPE GYRO COMPASS	DEPALMA TEWARI LYOMATA ASPDN	LAMAS	MEMORY OF WATER

10-1

Notes

(1) Guenther Wachsmuth, *The Etheric Formative Forces in Cosmos, Earth and Man* (Anthroposophical Publishing, London, 1932).

(2) Op. cit., pg. 9.

(3) Claude Bragdon, *Four Dimensional Vistas* (Knopf, New York, 1925), p. 51.

(4) Op. cit., pg. 124.

(5) Gustav Le Bon, *Evolution of Matter* (Walter Scott, New York, 1907), p. 90.

(6) Dmitri Mendelieff, *A Chemical Conception of the Ether* (Longmans Green, London, 1904).

(7) Le Bon, op. cit., pg. 33.

(8) C. Howard Hinton, *The Fourth Dimension* (Swan Sonnenscher, London, 1906), p. 230.

(9) Op. cit., pg. 230.

(10) Le Bon, op. cit., p. 97.

(11) John White and Stanley Krippner, *Future Science* (Doubleday, New York, 1977). Article by William A. Tiller, "The Positive and Negative Space/Time Frames as Conjugative Systems,' p. 257.

(12) Wachsmuth, op. cit.

(13) Michael Schiff, *The Memory of Water: Homeopathy and the Battle of Ideas in the Hew Science* (Thorsons, London, 1994).

(14) Trevor James Constable, *The Loom of the Future* (Borderlands, Bayside, 1996).

(15) David Morehouse, *Psychic Warrior: Inside the CIA's STARGATE Program* (St. Martins Press, New York, 1996).

(16) Peter Kelder, *Ancient Secret of the Fountain of Youth* (Harbor Press, Gig Harbor, WA 1989), p. 11.

Chapter 11
Final Observations

The etheric technological outline we have scanned is here and now. The barriers against immediate implementation are political, social, legal, intellectual and financial, and are also here and now. Almost all who gain from or have an interest in the contemporary physical paradigm in government, science, academe, and business constitute a more or less potentially uncooperative and often hostile barrier.

Knowledge of the new science paradigm and its technical potential is an esoteric knowledge limited to those who have broken the confines of intellectual inertia to visualize the future without the restrictions of the past. The famous science fiction writer Arthur C. Clarke is one who has openly stated initial hesitation, then acceptance:

"After initial skepticism I have now seen so many positive reports from highly respected organizations that there can be no further doubt that excess energy is being produced by some previously unknown process." (Letter to Vice President Gore published in *Cold Fusion*, May 1994).

The new paradigm is not widely known because our 3-space oriented academic science and media establishment is notably reluctant to understand or even to consider concepts and observations inconsistent with the contemporary view of the universe.

A 3-space world finds it hard to visualize a 4-space world, much as the hypothetical 2-space flatland observer would find it difficult to imagine a 3-space world. However, the view from 4-space is sharp and clear. The limitations of 3-space technology are highlighted. The enormous cost of pollution and the ineffective costly unwieldy regulatory method used to abate pollution is an unnecessary burden.

The cost of limited energy is an economic barrier to world standards of living. The disaster of desertification, the cost of natural disasters and inability to handle the multiple disease failures of modem medicine—all are crystal clear from the vantage point of 4-space and its dramatic solutions.

Yet the bounties offered by 4-space gated technologies are by no means universally welcome. A comment made by almost all those who have explored 4-space is the hostility and emotional distress created by 4-space ideas and interpretations. How far this emotional distress can be stretched is reflected in books like Gary Taubes' *Bad Science: The Short Life and Weird Times of Cold Fusion*. This book, written in 1993, describes the failure of cold fusion and the congenital inadequacies of cold fusion researchers. A most amazing volume, an ideal example of how people see what they want to see, and dismiss what is inconvenient for their world view. Of course, as we write in 1997, four years later, there is no question about the success of the cold fusion revolution. Yet, presumably, Gary Taubes is still editor of *Scientific American*. What this book really demonstrates is the weird naivety and long life of otherwise sensible scientific writers. Why? Simply because of the more than obvious threat to past investment in knowledge and finance in a 3-space world.

Contemporary physics is a tightly structured mathematical statement which cannot absorb or tolerate modifying input. Add new observations, modify one law and the entire mathematical construct collapses. The orthodox physicist is eminently correct in denying over unity, for example. Over unity cannot exist in three space. Orthodox pharmacology is correct in denying homeopathy where efficacy increases with dilution. Homeopathy is a 4-space technology and Avogadro's Limit does not apply in 4-space. Observers bound by the time and space phenomena of 3-space physical dimension will refute remote viewing and a 3-space bound Defense Intelligence Agency will never understand its parameters, especially if it uses Black Arts practitioners.

The mass media is a massive barrier. The new science has its own journals but these are low-circulation publications oriented towards workers in the field, and they do not reach the public at large. The mass circulation journals are controlled by those with deep investment in the contemporary system. Even worse, the editors and writers of science journals are locked into the 3-space view of the world. An excellent example is former editor of *Nature*, John Maddox, who slashed at every 4-space discovery that surfaced, e.g., morphogenetic fields, cold fusion, the second curve in homeopathy—all came under Maddox's withering, and sometimes dishonest, hatchet. This suggests that Thomas Kuhn in *Structure of Scientific Revolutions* has a valid argument. The new will supplant the old. The new will not be absorbed by the old. This suggests that articles and conferences with themes attempting to merge the old and the new have limited value although they do bring attention to the stark differences. In fact, given the absurd emotional distress emoted by John Maddox, Gary Taubes and many similar scientific writers, it might be safer to keep the various adherents separated.

In Diagram 10-1 we portray major 3-space anomalies rejected or ignored by conventional scientists but where objective evidence is known. These do not conform however to 3-space laws and definitions and every explanation from fraud to aberrant observations is used to push the evidence to one side. Why? Because if accepted they are inconsistent with 3-space laws and require the structure to be re thought and adjusted.

Rather than try to absorb or look for an explanation, orthodoxy prefers to shunt to one side to preserve a unified fabric of human made law. But practical 3-space cannot explain these anomalies—they have no causal explanation within 3-space. Nor has anyone apparently thought of grouping the anomalies and looking for a common causal origin in 4-space as we do here. These 3-space anomalies are actually the representation or the effects of 4-space energy in a 3-space world. *They are effects with a causal origin outside 3-Space.* They cannot be explained in the physical materialist paradigm because only the effect is seen and that needs a transducer, i.e., a device or system that will enable the effect to be identified in 3-space. The cause is in 4-space. One sees effects but no causes in 3-space.

This leads to the conclusion that those who search for some linking or merging of the two dimensions do not understand the cause and effect duality. One cannot merge cause and effect. To attempt to merge the two dimensions is impractical and will lead to confusion. Just look at what happens when orthodox scientists attempt to view these 4-space phenomena...they say flatly impossible and refuse to look further.

Moray B King in *Tapping the Zero Point Energy* (p. 13) illustrates with a diagram this distinction. A "flatland" slot represents 3-space (its width determined by Plancks constant) and surrounded by zero point energy (vacuum energy). "This," says King, "is not a passive system but actually is a manifestation of an energy flux passing through our space orthogonally from higher dimensions." (p. l2)

Orthodoxy does recognize the existence of zero point energy but argues it cannot be utilized because of incoherence. The energy is random and according to the Second Law of Thermodynamics must always remain random. However, this law is manmade and only applies within 3-space. In fact, a Russian scientist named Ilya Prigogine has expanded the laws of thermodynamics to show that systems can theoretically evolve from randomness towards order.

Whatever the predictions of 3-space laws, the fact is that we have devices that facilitate manifestations from 4-space, and in Diagram 10-1 we illustrate anomalies and their associated gating mechanisms. These are a place tested 4-space technologies drawing on 4-space energy and not conforming to 3-space man-made laws. While we cannot yet identify and describe the transducers in a unified way there is one 4-space technology that appears to include several transducers gating elements. The Viktor Schauberger cyclone turbine is a spiral section turbine which generates diamagnetism through a created vacuum. Diamagnetism is also common to human blood and breath, homeopathy and weather engineering.

In the Reich-DeMeo version of weather engineering the etheric force is actually termed orgone or life energy which is also diamagnetic. This suggests that diamagnetism, pretty much ignored in modem science and known as the repulsive force, needs to be investigated for its role in the etheric force.

Passing from the science discovery phase to the economic application phase we find that economic projections would be premature until some shaking out of the numerous devices has been seen.

The major technology shift is to energy and over unity generation. Over unity is impossible in 3-space but sooner or later an effective over unity device will become available whether using cold fusion or N-machines. From the economic viewpoint it will be more efficient than any existing 3-space generation system. Not entirely free energy but the fuel itself will be free, leaving capital and maintenance costs to be covered.

Apparently all over unity devices are low voltage DC, whereas our existing electricity generation systems are high voltage AC. This means the era of massive central generating stations with large economies of scale is over and the grid system becomes superfluous. This movement will be aided by the accumulation of evidence on electromagnetic pollution and adverse health effects.

The economic uses of over unity systems will be in standalone in its requiring only access to water. Thus, the factors entering into location of population and economic activity shift greatly. If there is pure water then any activity can be located anywhere in the world, *almost with equal advantage*. Combine this advantage with reversal of desertification using DeMeo techniques and one can easily see shifting world population and standards of living.

But here we come to another road block. Academic economics is as sterile as academic physics. There never has been any deep understanding of technology in the academic economics world, even though we have just emerged from a two century industrialization process. The emphasis is on sterile mathematical manipulation much like orthodox physics. Paradoxically, mathematics is a useful device in understanding the geometric nature of dimensions but when applied to economics becomes as sterile as dust with an almost medieval quality.

Fortunately, the world has emerged from its love affair with planning ("planned chaos," as Ludwig von Mises said) and now sees the usefulness of a market mechanism. This decentralized individual choice motivated economics fits neatly with the new science need. Whatever the final configuration of device used for over unity standalone units there are two guidelines:

The technical unknowns have to be solved leaving a full range of technical solutions for the application of market forces. It would be a disaster to fasten on an early technical solution only to find two or three decades later that a more technically efficient solution exists. This suggests the early and urgent investment of funds into workbench activity.

Market forces are the most efficient guide to construction and location of these units and also to sort out the least efficient.

Unfortunately, the excesses of 20th century capitalism have given the word "profit" a bad name in some areas. In fact, profit is a vital guide to efficiency and a vital component of economic efficiency. On the other hand, the apparent role of intent in 4-space suggests that profit will be limited to its guidance role, not including acquisitive and manipulative roles.

These rules are essential for an efficient utilization of the new processes. In brief, government should assist, but not determine, development. Hold in mind the dramatic failures of Soviet technology and the ongoing example of a confused U.S. Department of Energy.

What of the future? Be assured of one conclusion. The new science paradigm is not going to evaporate, disappear or be manipulated, suppressed or destroyed. At some time it will become the prevailing paradigm, but probably only after titanic struggles. The new will supplant the old sooner or later. Then we shall find that the future of the futurists is not the survivor. Professional futurists today extrapolate the present to find the future. They see super electronic doo dads, mechanical robots and genetic manipulation. They take the mechanical physical to the extreme and call it "the future." They may have thoroughly explored the possibilities of 3-space but they do not enter the realm of 4-space.

Why build multi-billion dollar space ships when a remote viewer can make the space trip as a daily routine? Where is the incentive to push the electronic frontier if electronic pollution threatens human survival and certainly health? This is without considering the absurdity of chemical cloud seeding and chemical fertilizer agriculture or the side effects of genetic engineering.

And of course 3-space cannot understand the human elements of the puzzle relating to consciousness and spirit. Today the awareness of 4-space is scattered and almost haphazard. There are unquestionably many who understand. Our guess is several tens of thousands of writers, scientists and intelligent laypeople are versed in one or several aspects. There are publications available. There are organizations and small groups working jointly to common ends within the new science. Very slowly the knowledge and awareness spreads. Yet the most successful 4-space technology, etheric weather engineering, is barely known. Another technology, Browns Gas, was seized and looted by a predator before it even got off the ground.

A likely route for development is through voluntary communities, shared values, and principles, especially as intent is an ingredient for 4-space success. Presumably, a joint intent will have a significant influence. In fact, early efforts were in such communities as Methernita in

Switzerland, Stelle, and Sunburst in the United States. In coming decades the emphasis will be on understanding and education, a widening of horizons, a deepening of appreciation.

The single factor likely to spur public interest is adoption of a device like cold fusion, or Browns Gas, or an etheric weather engineering program or remote viewing. These are dramatic advances. Most people are practical. They don't ask whether something conforms to the Laws of Thermodynamics. They don't care. They ask simply, "does it work?"

We already have transducers like gyroscopes and stabilizers but their benefits are not obvious to the layperson. A standalone heating device, a vehicle that runs on water, removal of atomic waste with Browns Gas, a drought relieved by weather guns. These are dramatic and direct. Once these are achieved public response will drive the hostile elements to the wall. Public awareness opens the door to politicians and the media.

Finally a word of caution. The emerging science has extraordinary implications. The entire world belief structure is now in process of change, it may take several centuries for the forces to play out.

An industrial era based on atomic and fossil fuels is finished. It will be replaced by a more powerful non-polluting, almost cost free technological system. Most will benefit, some will not. Those who will lose are those whose spirit is locked into the material objectives of power, control and ego stimulation. It is unrealistic to assume that the power groups will willingly surrender their material security blankets.

Our world "leaders" see success in terms of political power, in terms of control and ego enhancement. And they go trotting around the world to meet in pompous conferences allegedly in search of peace but more realistically what they have done is direct the bulk of the world's technology to war, corruption and destruction.

The revolution that ejects these elements will be a quiet revolution, what Laszlo terms "the whispering pond." Remember the Soviet col lapse came and went almost without a whimper. Russians withdrew their belief in the system. Anonymously, the fiction of the State was subverted and overthrown.

The revolution is from within: within the mind, within the individual, within the family and community. A revolution of the spirit and morality.

Appendix A
The Life Force: A Comprehensive Listing of Discoveries

Name of Force	Found In	Source
Manifesting force	Cosmology	Bayley
Life force	Biology	Vogel
Apergy (Keel)	Physical	Keely
Magnetoelectric	Physical	Tiller
Biological electrons/biotrons	Biology	Budwig
Fundamental energy/		
Universal energy	Biology	De La Warr
Universion	Biology	G. Lakhovsky
Shakti	Religion	Hindu
Mitogenetic radiation	Biology	A.Gurwitsch
Thought	Cosmology	James Jeans
Psi-fields	Psychology	W.G. Roll
Quanta energy packets	Physics	G.D. Wasserman
Anima mundi/cosmic		
Soul	Cosmology	Gustav T. Fechner
Holy Spirit	Religion	Christianity
Entelechy	Biology	Hans Driesch
Visual ray/eye beam	Biology	Lehrs
Ether/primary energy	Meteorology	Constable

Name of Force	Found In	Source
Ether gas	Chemistry	Mendelieff
Negative entropy	Physics	E. Schroedinger
Etherion	Chemistry	Brush
Zero point energy	Physics	Moray King
The Great Ultimate	Cosmology	Chu His (China)
Telesma Trimegistos	Occult	Hermes
Life rays	Biology	Crile
L-field/fields of life	Biology	HS Burr
Primordial field	Physics	Charles Leach
Orgone	Physics	Wilhelm Reich
Od/odic/odyll	Psychology	Baron von Reichenbach
Munia	Occult	Paracelsus
Archeus (magnus magnum)		J.B. von Helmont
vis Medicatrix naturae	Medicine	Hippocrates
Vital principle	Philosophy	Regnano
Spiritus	Occult	Robert Fludd
	Religion	Father Kircher
	Religion	V.G. Gassner
Magnetic fluid	Religion	Swedenborg
Animal magnetism	Biology	Franz Mesmer

Name of Force	Found In	Source
Animal magnetism	Biology	Dr. D'Elson
	Medicine	Dr. Escalente
Psychic fluid	Psychology	Emil Boirac
Nervous ether	Psychology	Dr. Richardson
Serpent power	Religion	Michel
Terrestrial magnetism	Physics	J.A. Fleming
Gestaltung	Cosmology	
Life ether	Cosmology	Steiner
Etheric Body	Cosmology	Steiner
Natural Energy	Archeology	
Terrestrial spirit		
Solar Spark		
Sun force	Archeology	Olive Pixley
Imagination of nature	Occult	Eliphas Levi
Astral light	Occult	Blavatsky
Spiritual essence	Biology	Goethe
Paraelectricity	Medicine	Ambrose Worral
Cosmic energy	Medicine	Harry Edwards
Animal electricity/		
life force	Biology	Luigo Galvani
Libido energy	Psychology	Sigmund Freud
Vital force	Medicine	Acupuncture
Bio-energetic energy	Biology	Alexander Lowen

Name of Force	Found In	Source
Effluence	Biology	Pietro
Life beams	Occult	Robert Fludd
Life force	Biology	Aldini
La fore vitale	Biology	Dr. Hippolyte
Prana	Religion	Hindu
Chi	Religion	China
Ki	Religion	Japan
Rauch	Religion	Hebrew
Nous	Philosophy	Plato
Yesod	Occult	Cabbalists
Formative cause	Philosophy	Aristotle
Pneuma	Philosophy	Erasistratus
Wodan	Philosophy	German
Facultas	Formatrix	Galen
Baraka	Religion	Sufis
Mana	Religion	Kahuna
Anima Mundi		Avicenne
Huaca		Peru
Orenda-Oki	Religion	Iroquois Indians
Manita	Religion	Algonquin Indians
Wahanda-Wahan	Religion	Sioux Indians
Sila	Religion	Eskimo

Name of Force	Found In	Source
Megbe	Religion	Buri pygmies
Elima	Religion	Congo pygmies
Rivu	Religion	Kalahari bush
Munga	Religion	Sudan
Wong	Religion	Gold Coast
Mulungu	Religion	Yaos
Ngai	Religion	Masai
Njom	Religion	Ekoi
Godhead	Religion	Trevelyan
Ayik	Religion	Elgonyi
Arunquiltha	Religion	Elgonyi
Churingi	Religion	Aborigines
Zego	Religion	Malaysia
Badi	Religion	Malaysia
Labuni	Religion	Gerlarin, NG
Ari, Hau	Religion	Penape, Pacific
Kasinga, Kalit	Religion	Palau, Pacific
Tondi	Religion	Balaka, Pacific
Ama	Religion	Maori, NZ
Andrimanitra	Religion	Malagasay
Amit	Religion	Kasule, Pacific
Tabi	Religion	Pacific

Name of Force	Found In	Source
Facultas formatrix	Physics	Johannes Kepler
Ectoplasm	Occult	Charles Richet
Vril		H. Bulwer-Lytton
Motor force	Physics	J.W. Keely
Neuric energy	Psychology	Dr. P. Barety
Neuricidad, Neural radiant energy		
It		Georg Groddeck
Libido	Psychology	S. Freud
Etheric formative forces	Cosmology	R. Steiner
N-Rays	Psychology	Blondlot
Magnotosim	Physics	A. Wendler
Hermic energy	Psychology	Wm. McDougall
Elan vital	Biology	Henri Bergson
Formative energy		Paul Kammerer
Vital magnetism		Charles Littlefield
Etheric force		Radaesthesists
Mitogentic radiation	Biology	Gurewitsch
Eloptic energy		T.G. Heironymous
Bioplasma	Biology	V.S. Grischenko
Psi faculty	Psychology	J.B. Rine
X-force	Medicine	E. Eeman
Soul of universe	Religion	Gustav Stromberg

Name of Force	Found In	Source
Dielectric biocosmic	Psychology	Oscar Brunler
Synchronicity	Psychology	Carl Gustav Jung
Psychotronic energy	Psychology	Robert Paulita
Psionics	Psychology	J.W. Campbell
Quasielectrostatic fields	Biology	Herny Morgenthau
Synergy	Psychology	Abraham Maslow
Psi plasma, Inergy	Psychology	Andreya Puharich
Primary perception	Psychology	Clive Backster
Biomagnetism	Biology	De La Warr
Time	Cosmology	Nikolai Kozyrev
Integrative tendency		Arthur Koestler
Anamorphosis		Ludwig von Bertalanity
Unitary principle		L.L. Whyte
Synergy	Architecture	Buckminster Fuller
X-Factor		Colin Wilson
Noetic energy		Charles Muses
Ra	Religion	Egypt
Vitic		Barnes
Ether-fifth element	Philosophy	Aristotle
N-Strahlen	Physics	Geffken

Name of Force	Found In	Source
Mitogentic radiation	Biology	Otto Rahn
N-Rays	Physics	De Lepiney
N-Rays	Physics	Cleaves
All creating force		C. Dietrich
Anthropoflux R		J.L. Farny
Life Fields	Biology	John Cage
Biofield	Biology	Buryl Payne
Biomneter	Medical	Adamenko
Nervous fluid	Biology	Mead
Energy field	Biology	Lowen & Pierrakos
Etheric vibration		Kedzie
Protoplasm		H.O. Ward
Pulse of the universe	Physics	Hallacy
Probability amplitudes	Physics	von Weizacker
Synergy	Physics	Lawrence Beyman
Shadow energy	Physics	Dr. Shiuji Inomata
Virtual energy		
Pan psychist consciousness		
Voidyon	Religion	Buddhism
Tachyon	Physics	Hans Nieper
Ether radiations energy	Physics	Dr. Joseph Matthias

Name of Force	Found In	Source
Magnetic fluid	Medicine	
Electron sea	Physics	P. Dirac
Lines of the world	Mythology	Yaqui Indians
Etheric web field	Occult	Caodai
Magnetic energy	Physics	Starr-White
Electromagnetic energy	Physics	Tromp
Human atmosphere	Physics	W.J. Kilner

Source: John White and Stanley Krippner, *Future Science* (Anchor Books, New York, 1977). The original list of about 100 discoveries is from White and Krippner. This was supplemented with additional research to create the present list of nearly 200 discoveries.

Appendix B

Sources of Information

Orgone Biophysical Research Laboratories Inc. Director, Dr. James DeMeo. Greensprings Center, P.O. Box 1148, Ashland, OR 97520; phone/fax: (541)552-0118; website: http://id.mind.net/community/orgonelab/index.htm. Catalog, numerous publications, conferences and lectures.

Planetary Association For Clean Energy (PACE). Director, Dr. Andrew Michrowski. 100 Bronson Ave., Suite 1001, Ottawa, Ont., Canada K1R 6G8. Long established reputable source for information. Webpage: http://www.energie.keng.de/-pace.

International Society For the Study of Subtle Energies and Energy Medicine. 356 Goldoc Circle, Golden, CO 80403, fax (303) 279-3539, website: http://www.vitalenergy.com/ISSSEEM/. Publish a journal entitled *Subtle Energies.* The title says it all. Now in its seventh year of publication.

Delta Spectrum Research. Route 3, Box 158-A, Inola, OK 74036. Vibration research institutes and laboratories. Strong on work of John Keely and acoustics. Excellent technical material.